Monjit Roy
Ajoy Hatibaruah

The Formula for Growth

Monjit Roy
Ajoy Hatibaruah

The Formula for Growth

Applying Mathematical Models and Psychology to Expand Your Business

LAP LAMBERT Academic Publishing

Imprint
Any brand names and product names mentioned in this book are subject to trademark, brand or patent protection and are trademarks or registered trademarks of their respective holders. The use of brand names, product names, common names, trade names, product descriptions etc. even without a particular marking in this work is in no way to be construed to mean that such names may be regarded as unrestricted in respect of trademark and brand protection legislation and could thus be used by anyone.

Cover image: www.ingimage.com

Publisher:
LAP LAMBERT Academic Publishing
is a trademark of
Dodo Books Indian Ocean Ltd. and OmniScriptum S.R.L publishing group

120 High Road, East Finchley, London, N2 9ED, United Kingdom
Str. Armeneasca 28/1, office 1, Chisinau MD-2012, Republic of Moldova, Europe
Managing Directors: Ieva Konstantinova, Victoria Ursu
info@omniscriptum.com

Printed at: see last page
ISBN: 978-620-8-43974-3

PREFACE

In today's fast-paced, interconnected world, businesses face a constant stream of challenges and opportunities that demand swift, calculated responses. Growth is no longer just about expansion or reaching a target—it is about creating a sustainable, adaptable model that can thrive amid changing consumer behaviors, technological advancements, and evolving market dynamics. To succeed in this landscape, business leaders must adopt a multi-faceted approach that blends the power of **mathematical analysis**, **data-driven decision-making**, and a deep understanding of **psychological principles**. This combination forms the backbone of a business strategy that is not only effective in the present but also resilient enough to withstand future disruptions.

This book is designed to equip entrepreneurs, managers, and leaders with the tools to build and sustain growth in their businesses. It explores the intersection of **math**, **data**, and **psychology**—three critical forces that, when integrated effectively, drive informed decision-making, optimize customer engagement, and ensure long-term success. While many business books focus on either the analytical or the human side of growth, this book takes a holistic approach, showing how these seemingly distinct elements work together to fuel innovation, adaptability, and competitive advantage.

The journey begins with a deep dive into the **psychological aspects** of leadership and team dynamics, recognizing that successful teams and companies are built on an understanding of human behavior. From there, we explore how **data** can inform decisions, enhance customer experiences, and predict future trends. In parallel, we discuss the role of **mathematical models**—such as predictive analytics and optimization techniques—that can transform raw data into actionable insights. By weaving these elements together, this book provides a comprehensive framework for businesses to not only survive but thrive in an increasingly complex business environment.

Throughout the chapters, we highlight how businesses of all sizes—from startups to large corporations—can harness the power of this triad of disciplines to address common growth challenges, scale across diverse markets, and foster cultures of innovation and collaboration. We also include case studies and practical examples of companies that have successfully implemented these strategies, offering real-world evidence of their impact on growth and sustainability.

The central message of this book is clear: growth is not a linear path. It requires constant learning, adapting, and innovating. By understanding and applying the principles outlined in these pages, business leaders can create models that are not only growth-driven but also resilient and sustainable in the long term.

In the chapters that follow, we will explore how businesses can utilize **data**, **psychology**, and **mathematics** to address immediate challenges and position themselves for future success. Whether you are seeking to improve your team's performance, enhance customer satisfaction, scale across multiple markets, or simply stay ahead of the competition, the principles in this book will provide a robust foundation for your journey. It is my hope that the insights presented here inspire you to take bold, calculated steps toward building a growth-oriented business that will stand the test of time.

Dr. Monjit Roy, an accomplished academic and professional, has 16 years of experience in teaching, 7 years of research, and leadership roles across schools, colleges, and universities. He holds an M.Ed, PhD, and multiple certifications in various fields such as computer applications, digital marketing, and mutual fund trading. Currently serving as an Assistant Professor and Research Guide at St. Joseph University, he has been involved in several research initiatives, seminars, and workshops. **Dr. Roy** has contributed significantly to academic development, acting as a resource person and guiding research scholars. With a diverse background in education, business, and leadership, he continues to make impactful contributions to both his institution and community. He is also an active member of several global academic networks and participates in various professional development programs.

Second Author

Dr. Ajoy Hatibaruah, Assistant Professor, Department of Mathematics, presently working in St. Joseph University, Nagaland. He has a teaching experience of 4 years. He has published many research papers in various National & international journals and attended many Seminar/Workshop/conferences. He holds Certification in computer Application.

Abstract

This book provides a comprehensive exploration of how **mathematical strategies**, **data-driven insights**, and **psychological principles** can be integrated to fuel business growth and development in the modern, fast-paced market. It examines the critical role these elements play in shaping long-term, sustainable growth strategies and addresses the challenges businesses face as they scale and adapt to ever-changing environments. By diving into predictive models, big data analytics, and behavioral psychology, the book illustrates how companies can optimize operations, anticipate market trends, and deepen their understanding of consumer behavior to make informed decisions.

Key themes explored include the application of **data analysis** to improve operational efficiency and customer engagement, the use of **psychological insights** to enhance marketing strategies and foster customer loyalty, and the importance of building **growth-oriented business models** that can thrive in both stable and volatile markets. The book also covers emerging trends in **technology**, **consumer behavior**, and **market dynamics**, offering strategies for businesses to remain competitive and adaptable in the future.

Through detailed frameworks, real-world case studies, and actionable strategies, this book equips readers with the tools and insights necessary to construct a growth strategy that is not only innovative but also resilient and scalable. It emphasizes the importance of continuous improvement, the value of psychological understanding in customer interactions, and the role of data in shaping decision-making processes. Ultimately, the book encourages readers to apply these principles in order to drive sustainable growth, navigate future challenges, and build successful businesses that can adapt and thrive over the long term.

ACKNOWLEDGMENTS

I would like to express my heartfelt gratitude to God for blessing me with countless opportunities. I am deeply thankful to my parents, who worked tirelessly and invested so much in my education, ensuring that I am now equipped to face the world confidently, regardless of the circumstances. To my friends, I thank you for your jealousy and criticism, for they have only made me stronger and more determined to excel in all aspects of life and work. Lastly, I am grateful to myself for putting in the effort to overcome life's challenges and embrace its ups and downs with confidence.

A special thanks to St. Joseph University, Nagaland for providing me the opportunity to serve as an assistant professor, a role that I cherish and will continue to grow in.

"The Formula for Growth: Applying Mathematical Models and Psychology to Expand Your Business"

CONTENTS

Chapter-1: Introduction

In today's competitive business environment, growth is the ultimate goal for any organization. "The Formula for Growth" explores how businesses can harness the power of mathematical models and psychological principles to drive sustainable expansion. By combining data-driven strategies with an understanding of human behavior, companies can uncover new opportunities, optimize their operations, and achieve long-term success. This book delves into the practical application of these fields to help business leaders make informed decisions and fuel their growth.

Chapter-2: Understanding the Intersection of Data, Math, and Psychology in Business Growth

In today's fast-paced, constantly evolving, and highly competitive business landscape, traditional growth strategies—those that often rely on intuition, experience, and incremental improvements—are no longer enough to guarantee long-term success. While these elements have served businesses well in the past, the rapid shifts in consumer behavior, technological advancements, and market trends demand a more robust and adaptive approach. Businesses that cling to outdated methods risk falling behind, as they may lack the insights needed to respond effectively to emerging challenges and opportunities. In order to truly unlock sustainable growth, companies must embrace a more data-driven and scientifically informed approach, one that incorporates not only data analysis but also mathematical models and psychological insights. This integrated approach moves beyond simple observation, assumptions, or guesswork. It shifts the focus toward leveraging objective insights derived from data, mathematical precision, and an understanding of the psychological factors that influence consumer behavior.

Data analysis is a powerful tool that allows businesses to objectively assess their operations, consumer behaviors, and market dynamics. By gathering, processing, and analyzing data, businesses can identify trends, uncover opportunities, and make informed predictions about future behaviors. This data-driven approach enables organizations to move away from guesswork and base their decisions on solid, evidence-backed information. However, while data analysis provides a strong foundation for decision-making, it is not enough on its own to fully unlock a business's growth potential. This is where mathematical models come into play. By applying mathematical and statistical techniques to the data, businesses can create predictive models that not only help explain current trends but also forecast future outcomes. These models allow businesses to optimize their strategies, assess potential risks, and make smarter decisions about where to allocate resources, how to fine-tune marketing efforts, and how to plan for growth.

Equally important to this equation is the role of psychology, which often tends to be overlooked in traditional business growth strategies. Psychology plays a crucial role in understanding consumer behavior, which is often driven by emotions, cognitive biases, and subconscious motivations. By tapping into psychological insights, businesses can more effectively tailor their products, services, and marketing strategies to meet the needs and desires of their customers. This deep understanding of human behavior allows companies to connect with their target audience on a more profound level, creating stronger relationships, improving customer experiences, and driving greater customer loyalty. Additionally, psychology also influences how businesses interact with their internal stakeholders—such as employees, managers, and leadership teams. Organizational psychology can be a game-changer in improving employee productivity, morale, and creativity, all of which contribute to the overall success of the company.

The intersection of data analysis, mathematical models, and psychology represents the future of business growth strategies. By combining these three pillars, businesses are equipped to understand the underlying factors driving both consumer behavior and internal operations. They can make data-backed decisions, anticipate market shifts, and understand not only what customers are doing, but why they are doing it. This comprehensive approach enables organizations to create strategies that are more agile, adaptive, and, ultimately, successful. Rather than simply reacting to market forces, businesses can proactively shape their future by anticipating trends, optimizing customer experiences, and developing growth strategies based on a deeper understanding of human behavior and mathematical precision.

This integrated framework moves businesses beyond the limitations of traditional approaches. It provides the tools needed to thrive in today's complex and unpredictable market environment. As we explore the relationship between data, math, and psychology in this book, we will examine how businesses can harness these three powerful elements to drive growth, optimize performance, and build long-term success. We will look at how data-driven decisions can uncover new opportunities, how mathematical modeling can help guide future growth, and how psychology can enhance customer engagement and employee performance. The goal of this book is to equip you with the knowledge and tools to take your business to the next level by understanding and applying the principles that will shape the future of business growth.

By embracing the full potential of data analysis, mathematical models, and psychology, businesses can unlock new levels of insight and create more effective, targeted strategies for growth. This holistic approach is no longer a luxury for leading-edge companies—it has become an essential framework for survival and success in today's competitive, data-driven world. Whether you are a small startup or a large corporation, understanding how to combine these elements will provide you with a competitive edge and help you create a sustainable growth trajectory that is based on informed decisions, predictive accuracy, and deep psychological understanding. Through the exploration of these principles in this book, you will learn to harness the power of this intersection to make more effective decisions, build stronger connections with your customers, and ultimately grow your business in a meaningful and lasting way.

The Importance of Combining Data Analysis, Mathematical Models, and Psychology

The intersection of data analysis, mathematical models, and psychology forms a powerful and comprehensive framework for understanding, predicting, and ultimately driving business growth. Each of these elements plays a unique yet complementary role, and when combined, they enable businesses to make more informed, strategic decisions that lead to more sustainable and scalable growth. While data analysis provides a factual and objective foundation, mathematical models bring quantitative precision, and psychology offers a deeper understanding of human behavior—both internally within organizations and externally in the marketplace.

Data analysis serves as the bedrock of decision-making, providing businesses with insights derived from large sets of information. This information is gathered from various sources, such as customer interactions, sales data, market trends, and consumer behavior patterns. By analyzing these data points, organizations can identify key trends, predict future outcomes, and optimize their operations. However, while data analysis can highlight what is happening, it may not fully explain why something is occurring. This is where mathematical models come in.

Mathematical models take the raw data and apply quantitative techniques to analyze complex relationships between variables, enabling businesses to simulate various scenarios and predict future trends with greater accuracy. These models offer a structured way of interpreting data that allows for better decision-making in areas such as resource allocation, pricing strategies, demand forecasting, and risk management. They provide the "how" behind the insights that data analysis uncovers, transforming abstract data into actionable strategies. Mathematical models are particularly valuable when businesses need to predict the potential outcomes of different actions, helping them to minimize risks and maximize returns.

However, while data and mathematical models can provide valuable insights and predictions, they fall short without a human-centric understanding of behavior. This is where psychology plays a crucial role. Psychology, particularly consumer psychology, offers insights into why people make the decisions they do, how emotions and cognitive biases influence buying choices, and how individuals and groups behave in different contexts. Understanding the psychological drivers behind consumer behavior allows businesses to create more personalized and engaging experiences, whether through marketing campaigns, product design, or customer service strategies.

Moreover, psychology is essential in shaping the internal dynamics of organizations. Human behavior within a business—such as employee motivation, leadership effectiveness, and organizational culture—can significantly impact productivity, innovation, and overall success. By integrating psychological principles, businesses can foster environments that support collaboration, creativity, and high performance, leading to better outcomes both for employees and the organization as a whole.

When these three elements—data analysis, mathematical models, and psychology—are combined, they create a holistic approach to business strategy that is both grounded in evidence and sensitive to the nuances of human behavior. This integrated approach allows businesses to not only understand what is happening in the market and how it will unfold but also to anticipate how individuals will respond to different stimuli, messages, or products. By applying data and models with an understanding of psychology, companies can make decisions that are not only logically sound but also emotionally resonant with their target audience.

Furthermore, this combination empowers businesses to remain agile and adaptable in an increasingly complex and competitive environment. With data providing a real-time view of market conditions, mathematical models enabling precise forecasting, and psychology informing how to engage customers and employees effectively, businesses can stay ahead of trends, respond to challenges more quickly, and create stronger, longer-lasting connections with their audience.

The integration of data analysis, mathematical models, and psychology forms a robust framework for achieving business success. Each element contributes essential insights that, when combined, help organizations make better decisions, improve operations, and ultimately drive sustainable growth. By embracing this multidisciplinary approach, businesses can gain a competitive edge in a rapidly evolving marketplace and position themselves for long-term success.

Data Analysis: The Foundation of Business Decisions

At the core of any effective business strategy lies data analysis, which provides a factual, evidence-based foundation upon which decisions are made. In a world where data is abundant, understanding how to collect, process, and interpret this information has become crucial for businesses striving to remain competitive. Data analysis allows businesses to uncover valuable insights into a variety of factors such as market trends, consumer behavior, customer preferences, and operational performance. By analyzing historical data, companies can identify patterns and trends that help them better understand how their products or services perform and where improvements may be needed.

For instance, understanding which products are most popular among specific customer segments, or recognizing emerging trends in the market, can provide a business with the foresight to make more strategic decisions. These insights can highlight areas where demand is growing, or where shifts in consumer preferences are taking place, enabling companies to proactively adjust their strategies. With this knowledge, businesses can tailor their offerings to meet the evolving needs of the market, ensuring they stay relevant and appealing to their target audience.

Data also plays a crucial role in identifying inefficiencies and areas for optimization within a business. By examining operational data, companies can spot bottlenecks in their processes, discover underperforming areas, and find ways to improve resource allocation. Whether it's adjusting supply chain processes, refining inventory management, or enhancing customer service, data analysis provides the insights necessary to fine-tune operations. This helps reduce waste, improve productivity, and lower costs, ultimately leading to a more streamlined and efficient business model.

Additionally, data analysis enables businesses to reduce risk by providing a more informed view of potential challenges and opportunities. In the ever-changing business landscape, risks are inherent, but data-driven insights can help mitigate uncertainty. For example, analyzing sales trends and consumer sentiment can help businesses forecast future demand and adapt to market fluctuations. Predictive analytics, built on historical data, can also provide businesses with a proactive approach to addressing potential issues before they become significant problems.

Moreover, data empowers businesses to make smarter, more targeted decisions that maximize profitability while minimizing risk. By relying on hard evidence rather than gut feelings or assumptions, companies can ensure that their decisions are grounded in reality. These results in more effective marketing strategies, more personalized customer experiences, and more informed product development, all of which contribute to long-term business success. In this way, data analysis becomes not just a tool for understanding the present, but also a strategic asset for shaping the future.

In essence, data analysis transforms raw information into actionable insights that guide business decisions, providing a clear roadmap for growth. It helps companies understand what is working, what is not, and where opportunities lie, ensuring that every decision is backed by evidence and aligned with the company's long-term goals. With data at the heart of decision-making, businesses can navigate complexities, seize opportunities, and continuously adapt to the ever-changing market environment, positioning themselves for sustainable success.

Mathematical Models: Quantifying Relationships for Predictive Power

While data analysis provides essential insights, mathematical models enhance these insights by adding quantitative rigor. By applying statistical methods and algorithms, businesses can quantify relationships between various variables and predict future outcomes with much greater accuracy. Mathematical models, such as regression analysis, machine learning algorithms, or optimization techniques, can help businesses forecast everything from customer demand to supply chain needs, product performance, and market conditions.

For example, a business can use mathematical models to analyze the relationship between marketing spend and sales performance, allowing it to determine the optimal marketing budget to maximize return on investment (ROI). In finance, these models can help predict stock price movements or assess the risk associated with different investments. In operations, they can be used to streamline production processes, balance inventory, or predict resource needs. Mathematical models allow businesses to go beyond simple analysis of past trends by making data-driven projections about the future, helping organizations optimize strategies, allocate resources more effectively, and reduce the uncertainty inherent in decision-making. These models empower businesses to plan proactively rather than reactively, anticipating market changes and positioning themselves ahead of the competition.

Psychology: The Human Element of Decision-Making

While data and math provide a solid foundation for decision-making, psychology adds the critical human element. Business strategies that fail to consider psychological factors are likely to miss the mark in understanding why consumers make certain choices or how employees respond to organizational changes. Psychology helps businesses understand the motivations, needs, emotions, and behaviors of customers and employees alike.

In terms of customer behavior, psychological insights allow businesses to recognize that purchasing decisions are often influenced by subconscious factors such as emotions, cognitive biases, and social pressures. For instance, a customer's decision to buy a product may not be purely rational but could be driven by emotional appeal, fear of missing out (FOMO), or brand loyalty. Understanding these psychological drivers enables businesses to create more effective marketing campaigns, design products that resonate more deeply with customers, and develop services that meet their emotional as well as functional needs. By leveraging principles from consumer psychology, such as the scarcity effect or reciprocity, businesses can significantly enhance their ability to influence and persuade their target audience.

Psychology also plays a crucial role in shaping employee behavior and performance within an organization. Understanding how to motivate employees, foster innovation, and build a positive organizational culture can lead to better collaboration, higher productivity, and greater overall success. For example, principles from organizational psychology can guide leaders in creating a work environment that promotes creativity, supports employee well-being, and fosters trust and engagement. When businesses understand and apply psychological insights to both customer and employee interactions, they create a more harmonious and productive environment that drives growth on both fronts.

The Synergy of Data, Math, and Psychology

When combined, data analysis, mathematical models, and psychology form a powerful synergy that gives businesses a unique advantage. Data provides the raw material; mathematical models provide the tools for understanding and predicting behavior, while psychology brings the human insight needed to optimize customer interactions, employee engagement, and overall decision-making.

The power of this integration lies in its ability to generate holistic business strategies. With data-driven insights, businesses can understand the facts and figures that are driving their industry. Mathematical models help quantify these insights, giving businesses a roadmap for navigating future challenges and opportunities. Psychology adds the final layer, ensuring that the strategies implemented are not only logically sound but also emotionally and psychologically effective.

By combining these three disciplines, businesses are empowered to make more informed, precise, and human-centered decisions. This multidimensional approach allows organizations to create growth strategies that are both data-informed and people-centered—leading to higher customer satisfaction, increased loyalty, and a more motivated workforce. Whether it's predicting consumer trends, optimizing marketing campaigns, improving internal processes, or fostering a better organizational culture, the intersection of data analysis, mathematical models, and psychology offers the tools and insights necessary to propel businesses toward long-term success.

Why Traditional Business Growth Strategies May Fall Short

Traditional business growth strategies have often centered around familiar concepts like market expansion, operational efficiency, and cost-cutting measures. These approaches focus on scaling up operations, increasing market share, and streamlining processes to improve profitability. While these strategies can certainly yield short-term success, they tend to fall short in the long run because they do not adequately address the underlying factors that drive consumer behavior and business performance. In today's complex and rapidly changing business environment, relying solely on these conventional strategies is no longer enough to ensure sustained growth and success.

One of the major shortcomings of traditional growth strategies is their tendency to overlook the critical psychological and mathematical factors that influence both consumer choices and overall business operations. For example, traditional strategies often focus on increasing sales volume, reducing costs, or expanding into new markets without fully considering how consumer behavior or decision-making processes are evolving. Without understanding the psychological triggers behind customer decisions, businesses may miss opportunities to connect with their target audience on a deeper emotional or cognitive level. Products, services, and marketing campaigns that fail to resonate with customers' subconscious needs and desires are more likely to struggle, no matter how effective the operational strategies may be.

Similarly, traditional strategies often overlook the power of data and mathematical modeling in decision-making. In the past, business decisions were primarily based on experience, intuition, or historical trends. While this approach may have worked in more stable markets, the increasing availability of data and the complexity of modern markets require businesses to adopt a more data-driven approach. Without leveraging data to inform decisions, businesses risk operating on incomplete information or outdated assumptions. Market dynamics, customer behavior, and competitive landscapes can shift quickly, and relying solely on traditional methods means that businesses are likely to fall behind in an increasingly data-intensive and fast-paced world.

For instance, when a company relies purely on cost-cutting measures, they might overlook the long-term impact of reducing customer service quality or failing to innovate products. In such cases, while short-term profits might increase, customer loyalty and long-term growth can decline. Similarly, expanding into new markets without considering the nuances of local consumer behavior or market conditions could result in costly missteps. These traditional strategies, though effective in some cases, lack the foresight needed to navigate the dynamic nature of modern business environments.

Incorporating Mathematical Models and Psychological Insights

The missing ingredients in many traditional growth strategies are the integration of mathematical models and psychological insights. Mathematical models, based on data analysis, offer businesses the ability to predict future trends and outcomes with much greater accuracy. They allow businesses to test various scenarios, simulate outcomes, and optimize their strategies for better performance. For example, businesses can use predictive analytics to forecast customer demand, adjust supply chain management accordingly, or optimize marketing spend for maximum return on investment. These models not only provide a more scientific approach to decision-making but also ensure that businesses are better prepared for future challenges by identifying patterns and forecasting potential market shifts.

Psychology, meanwhile, adds a crucial human element to the equation. It helps businesses understand what drives consumer behavior, how emotions influence decision-making, and why customers sometimes act irrationally or unpredictably. Without incorporating psychological principles into their strategies, companies may struggle to design products, services, or marketing campaigns that truly resonate with their audience. For instance, an effective pricing strategy isn't just about cost-plus markup or competitive pricing; it also involves understanding how price influences perceptions of quality or how discounts may appeal to customers' psychological biases. Without tapping into these insights, businesses may miss the mark in their attempts to connect with customers and drive engagement.

Incorporating psychological insights into the growth equation ensures that businesses are not just reacting to market changes, but actively anticipating and shaping them. By understanding the emotions, motivations, and behavioral patterns that drive consumer decisions, companies can create more targeted and impactful strategies. Similarly, understanding how employees respond to different management approaches, organizational changes, and incentives can help businesses foster a more productive and innovative workforce. This adaptive mindset allows companies to stay ahead of market shifts, build stronger customer relationships, and create a sustainable competitive advantage.

Fostering Adaptability and Resilience

The integration of mathematical models and psychological insights into business growth strategies also fosters greater adaptability and resilience. Traditional approaches often focus on optimizing existing processes or replicating strategies that have worked in the past. While this can be effective in stable or predictable environments, it falls short in today's increasingly dynamic and volatile markets. In contrast, a strategy that incorporates data-driven insights and an understanding of psychological factors enables businesses to be more flexible in responding to external changes, customer demands, or emerging trends.

By anticipating market shifts, customer needs, and competitor actions through predictive models, businesses are better equipped to adjust their strategies proactively. Moreover, psychological insights into consumer behavior help businesses better understand and respond to changes in the marketplace, whether it's a shift in customer preferences, an emerging social trend, or the impact of new technologies. This ability to adapt quickly and strategically is critical in maintaining relevance and competitiveness in today's fast-changing business environment.

Additionally, fostering resilience through psychological understanding extends beyond consumer behavior to the internal culture of the organization. A company that understands the psychological needs and motivations of its employees is better able to build a resilient workforce that can navigate challenges, innovate, and drive growth even in times of uncertainty. Employee engagement, satisfaction, and motivation are critical factors that influence productivity, creativity, and organizational success. Companies that fail to recognize the importance of psychological factors in managing and supporting their workforce may find it difficult to retain top talent, foster innovation, or maintain a positive workplace culture, all of which are essential for long-term success.

While traditional business growth strategies have their merits, they often fall short in addressing the complexities and dynamics of the modern business landscape. By overlooking the psychological and mathematical factors that drive both consumer behavior and internal performance, businesses risk making decisions based on incomplete information and outdated assumptions. Incorporating data-driven insights, mathematical models, and psychological principles into the growth equation provides a more holistic and strategic approach to business success. This approach not only enables businesses to better understand and predict market trends but also helps them connect with customers on a deeper level and foster greater internal resilience. As markets become more unpredictable and competition more fierce, businesses that adopt this integrated strategy will be better positioned to thrive in an increasingly complex and dynamic environment.

Core Themes of the Book

This book delves deeply into the integration of three critical elements—data analysis, mathematical models, and psychology—and how their powerful intersection can drive business growth. In today's rapidly evolving market landscape, businesses can no longer rely solely on intuition or traditional strategies to sustain growth. To thrive, organizations must combine data-driven insights, mathematical precision, and psychological understanding into a holistic approach to growth. Each chapter of this book will explore how to leverage these elements effectively, providing you with the tools and knowledge to craft strategies that are both efficient and profoundly attuned to the needs, behaviors, and motivations of your customers and workforce.

The first core theme centers around **data analysis** as the foundation of modern business decision-making. Data is the currency of today's business world, offering rich insights into everything from customer preferences to market trends and operational performance. The book will highlight how to collect, process, and analyze data to uncover patterns and trends that drive smarter, more informed decisions. You will learn how to turn raw data into actionable insights, optimizing business processes, forecasting future trends, and reducing risks—ultimately positioning your organization for growth in an increasingly competitive environment.

The second key theme is the application of **mathematical models** to translate data into predictive insights and refined strategies. Through the use of sophisticated models, businesses can quantify relationships, simulate outcomes, and test various scenarios before implementing them in the real world. This theme will explore the power of mathematical models to provide precision in decision-making. Whether it's optimizing pricing strategies, improving resource allocation, or forecasting demand, mathematical models serve as a bridge between raw data and tangible business outcomes, enabling companies to make more confident, informed choices that drive performance.

The third, and perhaps most human-centered, theme involves **psychology**—understanding the behavior, motivations, and emotions of both customers and employees. While data and mathematical models provide a technical and quantitative foundation for business strategy, psychology offers crucial insights into why people behave the way they do. This theme will delve into the principles of consumer behavior, decision-making biases, and organizational psychology to help businesses create more personalized, engaging, and effective strategies. By understanding how people think and act, organizations can build stronger relationships with their customers and create workplace cultures that inspire innovation and high performance.

Throughout the book, these three themes—data analysis, mathematical models, and psychology—will be woven together to create a cohesive and integrated framework for business growth. By exploring how each element complements the others, you will gain a comprehensive understanding of how to apply them in unison to drive more effective business strategies. You will be equipped with actionable insights to better anticipate market shifts, create more engaging customer experiences, and foster a work environment that supports both individual and organizational success.

The core themes of this book highlight the power of combining data, mathematics, and psychology to create business strategies that are not only grounded in evidence and precision but also deeply attuned to the human factors that influence decisions and drive growth. By mastering the integration of these elements, you will be prepared to navigate the complexities of the modern business landscape and build a foundation for sustainable, long-term success.

The following core themes will be explored throughout the book:

Using Data to Uncover Insights and Guide Decision-Making

The first crucial step in developing an effective growth strategy understands how to harness the vast amounts of data available to you. Data analysis is the foundation of any modern business decision-making process, offering valuable insights into everything from customer preferences to market trends and operational performance. In today's competitive business environment, the ability to effectively use data is no longer a luxury—it is a necessity. This section will provide you with a comprehensive framework for gathering, analyzing, and interpreting data, enabling you to uncover meaningful insights that can guide your business decisions.

We will begin by exploring the various **data sources** available to businesses today, from customer interaction data to financial reports and market research. In an age where data is generated at an unprecedented rate, understanding how to collect relevant and high-quality data is key. This book will cover best practices for gathering both structured and unstructured data, ensuring you are capturing the right information to fuel decision-making. You will learn how to establish data collection processes, whether through customer surveys, transaction logs, or digital analytics tools, that ensure consistency and reliability.

Once the data is collected, the next step is **data analysis**. With the right tools and techniques, businesses can mine this raw data to identify patterns, trends, and anomalies that would otherwise remain hidden. The book will delve into a variety of analytical methods, from basic descriptive statistics to more advanced predictive and prescriptive analytics. You will learn how to use these methods to spot emerging trends, understand market dynamics, and evaluate the performance of products, services, and campaigns. Whether through segmentation analysis, sentiment analysis, or regression models, these analytical techniques will provide you with the insights needed to make informed, data-driven decisions.

The book will also focus on how to **interpret** the results of data analysis and translate them into actionable insights. While raw data may provide valuable information, it is the interpretation of that data that unlocks its true potential. We will discuss how to synthesize analytical findings to develop a deeper understanding of customer behaviors, operational efficiencies, and market conditions. With a clear understanding of what the data is telling you, businesses can move from insights to **strategies**—formulating actionable plans that will drive growth and improvement across all aspects of the organization.

In addition to guiding decision-making, data analysis also plays a critical role in **identifying opportunities** and optimizing operations for maximum efficiency. By continuously monitoring and analyzing key performance indicators (KPIs), businesses can spot areas of underperformance, inefficiencies, and gaps in the market. Data can highlight where new opportunities lie—whether that means tapping into untapped customer segments, improving customer retention, or fine-tuning product offerings. Armed with this information, businesses can fine-tune their strategies, allocate resources more effectively, and prioritize initiatives that will deliver the highest returns.

Finally, the book will cover **tools and technologies** that can assist in transforming raw data into actionable insights. From business intelligence software to machine learning algorithms, we will explore the many tools available to automate and streamline the data analysis process. These technologies can help businesses scale their data operations, identify trends faster, and gain more accurate insights, ultimately allowing for quicker decision-making and more effective execution.

Using data to uncover insights and guide decision-making is not just about gathering information; it is about understanding how to translate that information into meaningful, strategic actions. By developing a robust data analysis framework and mastering the tools and techniques needed to interpret and act on your findings, you can unlock new opportunities, optimize business operations, and make informed decisions that propel your organization toward sustainable growth and success.

Building Predictive Models to Optimize Business Performance

Once you have a solid understanding of your data, the next step is to apply mathematical models that can help you predict future trends and optimize business performance. Predictive models allow businesses to project outcomes with a higher degree of accuracy, guiding everything from inventory management to marketing spend, product development, and customer acquisition strategies. The power of predictive modeling lies in its ability to transform historical data into actionable forecasts that enable businesses to make proactive decisions rather than reactive ones.

In this section, we will explore a variety of **mathematical modeling techniques**, starting with the foundational method of **regression analysis**. Regression analysis is a powerful statistical tool used to understand relationships between variables. By analyzing historical data, businesses can identify correlations between different factors, such as customer demographics, purchasing behaviors, and external influences like economic conditions. These insights allow you to predict future outcomes based on existing trends, making it easier to plan for demand fluctuations, inventory requirements, and marketing efforts. Whether it's predicting sales volumes, customer churn, or product performance, regression analysis provides a straightforward yet effective way to model and forecast future scenarios.

As businesses look to become more sophisticated in their predictive capabilities, **machine learning algorithms** offer a more advanced approach to modeling. Unlike traditional methods, machine learning models learn from data by identifying patterns and improving their accuracy over time. In this section, we'll dive into different types of machine learning algorithms—such as supervised learning, unsupervised learning, and reinforcement learning—and explain how they can be applied to various business challenges. For instance, machine learning can be used to forecast customer purchasing behavior, segment customers for targeted marketing campaigns, or predict which products are likely to sell best in the coming months. By leveraging the power of machine learning, businesses can make predictions with greater precision and adapt their strategies based on continuously evolving data.

Another critical aspect of optimizing business performance through predictive models is the use of **optimization methods**. Optimization techniques focus on maximizing or minimizing specific outcomes, such as maximizing profit, minimizing waste, or optimizing resource allocation. For example, in the context of inventory management, optimization models can help businesses determine the ideal stock levels to meet customer demand without overstocking or running out of key products. These models take into account various constraints—such as budget limitations, lead times, and capacity—and help businesses make the most efficient use of their resources. In marketing, optimization models can guide decisions on the optimal allocation of advertising spend to achieve the highest return on investment (ROI).

By learning to use these predictive models, businesses can forecast customer behavior, optimize resource allocation, and reduce risks associated with uncertain business environments. Predictive models give organizations the foresight they need to act confidently, whether it's adjusting inventory levels before a seasonal surge, fine-tuning marketing strategies to attract new customers, or identifying potential bottlenecks in the production process before they become issues. The ability to predict outcomes based on data-driven insights reduces uncertainty and helps businesses move from a reactive to a proactive stance in managing day-to-day operations.

Furthermore, the use of predictive models extends beyond day-to-day operational decisions; it also plays a pivotal role in **strategic long-term planning**. By accurately forecasting future trends—such as customer preferences, market shifts, and industry developments—businesses can position themselves for sustainable success. Predictive models allow companies to anticipate challenges and opportunities, enabling them to develop strategies that are not only responsive to the current landscape but also resilient to future changes. For example, by using predictive analytics to forecast emerging trends in consumer behavior, businesses can stay ahead of the competition by innovating and adjusting their offerings before the market demands change.

Building predictive models is an essential step in optimizing business performance. These models allow businesses to move beyond intuition and subjective decision-making, providing a data-driven, quantitative approach to forecasting and strategy development. Whether through regression analysis, machine learning algorithms, or optimization techniques, predictive models enable businesses to anticipate future trends, optimize their operations, and mitigate risks in an ever-changing business environment. This data-driven, predictive approach will not only help you manage day-to-day operations more efficiently but also position your business for long-term success in an increasingly competitive market.

Applying Psychological Principles to Improve Customer Engagement and Loyalty

Understanding the psychological drivers behind customer behavior is crucial to developing a business strategy that goes beyond mere transactional relationships. In today's competitive market, simply offering a product or service is no longer enough to guarantee customer loyalty. To build long-lasting relationships, businesses must tap into the deeper emotional and cognitive factors that influence consumer decisions. This section will explore key psychological principles that shape consumer decision-making and demonstrate how businesses can apply these insights to create more engaging, personalized experiences that resonate with their audience.

One of the most powerful psychological drivers of consumer behavior is **emotional appeal.** Consumers often make purchasing decisions based on how they feel rather than on logical reasoning alone. Emotions can heavily influence choices, whether it's the joy of owning a new product, the sense of belonging to a community, or the relief of solving a problem. Successful brands often evoke strong emotional responses through their marketing and customer experiences. For instance, many companies use emotionally-driven advertising campaigns to create an emotional connection with their audience. By appealing to emotions such as happiness, fear, trust, or nostalgia, businesses can create experiences that resonate deeply with customers and leave a lasting impression.

Social influence is another powerful psychological principle that shapes consumer behavior. People are often influenced by the opinions, behaviors, and preferences of others. This can manifest in many forms, including word-of-mouth recommendations, social proof, and influencer endorsements. Consumers tend to trust the opinions of their peers more than advertisements, which is why reviews, testimonials, and social media interactions can significantly impact purchasing decisions. By understanding the power of social influence, businesses can strategically leverage social proof—such as showcasing positive customer reviews, testimonials, and user-generated content—to encourage potential customers to take the leap and make a purchase. Social influence also plays a crucial role in building brand communities, where customers feel connected to each other and the brand, enhancing customer loyalty and engagement.

Cognitive biases are another essential aspect of consumer decision-making. These mental shortcuts often lead customers to make decisions that are not entirely rational but are influenced by biases and heuristics. For example, the **anchoring bias** causes customers to focus heavily on the first piece of information they see, such as an initial price, and compare all subsequent information to it. The **scarcity effect**, where people place a higher value on items perceived as scarce, can also drive purchasing behavior. Recognizing and understanding these biases allows businesses to tailor their pricing strategies, promotional offers, and even product placement to align with how customers naturally process information. By leveraging cognitive biases in a thoughtful and ethical manner, businesses can influence decision-making in a way that leads to higher conversion rates and increased customer satisfaction.

Another powerful psychological principle is **the power of storytelling**. Humans have a deep connection to stories, and storytelling has been used for centuries to engage and influence others. In business, storytelling can be a powerful tool to connect with customers on an emotional level. By telling a compelling story about the brand, its values, or how a product solves a customer's problem, businesses can create a narrative that resonates with their audience. A well-crafted brand story not only makes a company more relatable but also builds trust and authenticity, which are key factors in fostering customer loyalty. Storytelling can be applied across various touchpoints, from advertising campaigns to website content and social media posts, making the customer journey more engaging and memorable.

By applying these psychological principles, businesses can create **personalized experiences** that cater to the unique emotional triggers and cognitive preferences of their customers. Personalization is key to standing out in today's crowded marketplace. By using data to understand individual customer preferences and tailoring communication, product recommendations, and offers, businesses can foster a deeper connection with their customers. For example, email marketing campaigns that address customers by name, recommend products based on past purchases, or provide exclusive offers create a sense of value and recognition, which drives engagement and increases the likelihood of repeat business.

When businesses focus on the emotional and psychological aspects of customer behavior, they can significantly improve **customer satisfaction** and **retention rates**. By meeting customers' emotional needs and making them feel understood, valued, and connected, businesses can create loyal customers who return time and again. This deeper level of engagement fosters trust and reduces the likelihood of customers switching to competitors, even when they offer similar products or services.

Understanding and applying psychological principles is key to creating a customer experience that goes beyond transactional exchanges and builds meaningful, long-term relationships. By leveraging emotional appeal, social influence, cognitive biases, and the power of storytelling, businesses can enhance customer satisfaction, improve retention rates, and foster deep brand loyalty. This section will equip you with the tools to craft strategies that not only attract customers but also keep them coming back for more, creating a loyal base that drives sustained growth for your business.

Understanding the Role of Behavioral Economics in Shaping Consumer Choices

Behavioral economics blends insights from psychology and economics to explain how consumers make decisions, often in ways that contradict traditional economic theories, which assume that individuals always act rationally and in their best financial interests. In reality, consumer choices are often influenced by cognitive biases, emotions, and psychological factors that lead to decisions that seem irrational or inconsistent with standard economic models. In this section, we will explore key concepts from behavioral economics—such as loss aversion, mental accounting, and framing effects—and examine how they shape consumer behavior. Understanding these psychological biases gives businesses a powerful toolkit for crafting marketing messages, pricing strategies, and product offerings that resonate with consumers' subconscious decision-making processes.

One of the most influential concepts in behavioral economics is **loss aversion**. This principle suggests that consumers are more motivated by the fear of losing something than by the desire to gain something of equal value. In other words, people tend to feel the pain of loss more acutely than the pleasure of an equivalent gain. For businesses, this means that potential customers may be more likely to take action when they perceive a loss or missed opportunity. For example, limited-time offers or "last chance" promotions tap into consumers' fear of missing out (FOMO) and can drive them to act quickly. Additionally, companies can frame discounts or bonuses in ways that highlight the potential loss if they don't take advantage of the offer, increasing the likelihood of a purchase.

Another key concept in behavioral economics is **mental accounting**, which refers to the tendency of consumers to treat money differently depending on its source or intended use. For instance, people might be more willing to spend "extra" money—such as a tax refund or a bonus—on indulgent purchases compared to money from their regular paycheck, even though the money is functionally equivalent. Businesses can use this insight to their advantage by framing prices and offers in a way that makes them feel like a "windfall" or "bonus" rather than an everyday expense. For example, offering a special "holiday" discount or exclusive deal for loyal customers can tap into this mental accounting bias, making customers feel like they are receiving something extra or special, thus encouraging more spending.

Framing effects are another significant behavioral bias that affects consumer decisions. The way information is presented (or "framed") can have a substantial impact on how people interpret it and the choices they make. For example, consumers are more likely to choose a product if it's presented as "90% fat-free" rather than "10% fat," even though both descriptions mean the same thing. This is because the positive framing (emphasizing what is present rather than what is absent) resonates more with consumers' preferences. Marketers can use framing to influence consumer perceptions and decisions by highlighting the benefits or advantages of a product, service, or offer, rather than focusing on potential downsides. For example, when promoting a product, it may be more effective to frame it in terms of how it will enhance the customer's life or solve their problems, rather than simply listing features or price.

In addition to framing, **anchoring** is another cognitive bias that businesses can leverage to influence consumer behavior. Anchoring occurs when consumers rely too heavily on the first piece of information they see when making a decision. This could be an initial price point, which then becomes the reference point for subsequent comparisons. For example, if a product is initially priced at $100 but is offered at a discount of $30, consumers may perceive the sale price of $70 as a great deal, even if the true value of the product doesn't justify the original price.

Marketers can use anchoring to their advantage by first showing a higher "original" price and then offering a significant discount, making the lower price seem more attractive.

Understanding **scarcity** and its impact on consumer decisions is another important insight from behavioral economics. The scarcity effect is the principle that consumers place a higher value on products or opportunities that are perceived as scarce or in limited supply. This is linked to the idea that people are often more motivated to avoid losing something than to gain it, a concept closely related to loss aversion. Creating a sense of urgency through "limited-time offers," "limited availability," or "only a few left in stock" can encourage customers to make a purchase, often more quickly than they otherwise would. By strategically communicating scarcity, businesses can increase the perceived value of their products and drive faster decision-making from customers.

By leveraging **behavioral economics**, businesses can better understand and anticipate how customers think and make decisions, even when those decisions seem illogical or irrational. Armed with this knowledge, businesses can craft more compelling offers, optimize pricing strategies, and design products or services that resonate with customers' subconscious decision-making processes. Whether through framing, mental accounting, loss aversion, or scarcity, these psychological biases can be harnessed to influence consumer behavior in ways that drive conversion rates and enhance customer loyalty.

Behavioral economics provides a rich understanding of the factors that shape consumer choices, often in ways that contradict traditional economic assumptions. By applying these principles, businesses can create marketing campaigns and strategies that tap into consumers' irrational but predictable behavior, ultimately guiding customers toward decisions that benefit both the consumer and the business. This section will equip you with practical tools to apply these insights to improve conversion rates, build stronger customer relationships, and ultimately increase your business's bottom line.

The Importance of Mindset and Organizational Psychology in Fostering Innovation and Employee Productivity

While much of the focus in business growth strategies tends to be on external factors such as consumer behavior and market dynamics, **internal factors**—especially organizational culture—are just as critical. A business's internal environment, including the mindset of its leadership and the psychological well-being of its employees, plays a significant role in determining its success. The attitudes, beliefs, and motivations within an organization often serve as the foundation for its overall performance. In this section, we will explore how cultivating the right mindset of growth, innovation, and resilience within your organization can lead to improved productivity, creativity, and long-term success. By understanding key concepts from **organizational psychology**, businesses can tap into their most valuable resource: their people.

Mindset is a foundational concept in this context. The mindset of both leadership and employees can profoundly influence a company's ability to innovate and remain competitive. Leaders who adopt a **growth mindset**—the belief that abilities and intelligence can be developed through effort and learning—are more likely to foster an environment where experimentation, problem-solving, and learning from failure are encouraged. This approach contrasts with a **fixed mindset**, which assumes that skills and talents are static. Leaders with a growth mindset promote adaptability, resilience, and the willingness to take calculated risks, which are essential qualities for fostering innovation.

In addition to leadership mindset, the overall **organizational mindset** plays a vital role in shaping the company culture. Organizations that prioritize continuous improvement, collaboration, and open communication create an atmosphere that supports creativity and the sharing of ideas. A culture that embraces failure as a stepping stone to success rather than something to be feared or avoided is crucial in encouraging employees to experiment and propose new ideas without the fear of negative consequences. This is especially important in fostering **innovation**—the ability to generate new ideas, products, or processes that can drive the business forward and set it apart from competitors.

Organizational psychology also plays a critical role in shaping the **motivation** of employees. Understanding what drives employees—whether it be intrinsic motivation (the internal desire to perform tasks because they are personally fulfilling) or extrinsic motivation (external rewards like bonuses or promotions)—can help businesses tailor their management strategies. Theories such as **Maslow's Hierarchy of Needs** and **Herzberg's Two-Factor Theory** provide frameworks for understanding how different needs (from basic security needs to the desire for recognition and self-actualization) affect employee motivation. By addressing these needs, businesses can create a more **motivating work environment**, where employees feel valued, recognized, and driven to contribute their best work.

Effective **leadership** is another key factor in fostering innovation and productivity. The leadership style adopted by managers can directly impact employee engagement, collaboration, and job satisfaction. Transformational leadership, which focuses on inspiring and motivating employees to exceed expectations through a shared vision and by offering personal support and encouragement, has been shown to foster higher levels of innovation. Leaders who provide **autonomy** and **empowerment** allow employees to take ownership of their work, which increases job satisfaction and leads to higher levels of creativity and productivity. On the other hand, authoritarian leadership styles, which focus on control and micromanagement, can stifle creativity and reduce employee engagement.

Another essential element in organizational psychology is **teamwork**. Successful businesses often thrive because they encourage collaboration and foster a team-oriented environment. Research in organizational psychology shows that teams with high levels of trust, communication, and diverse skills tend to outperform those with low cohesion. In high-performing teams, members contribute their unique strengths, share knowledge, and collaborate to solve complex problems. Leaders can cultivate these collaborative environments by providing opportunities for team-building activities, promoting cross-functional work, and fostering a culture where employees feel comfortable expressing their ideas and challenging the status quo. A collaborative mindset not only drives productivity but also enhances the ability to innovate by pooling diverse perspectives.

The **psychological well-being** of employees also has a direct impact on their performance. Organizations that prioritize the mental health and well-being of their employees tend to see higher levels of engagement, job satisfaction, and overall productivity. By offering resources for stress management, mental health support, and work-life balance, companies can help employees manage workplace pressures more effectively. Promoting a **positive work environment**, where employees feel supported, respected, and valued, encourages resilience and helps to reduce burnout. When employees feel psychologically safe, they are more likely to take initiative, share new ideas, and feel empowered to make decisions.

At the heart of fostering a positive organizational culture is the emphasis on **employee satisfaction**. Employee satisfaction goes beyond competitive salaries and benefits; it involves creating an environment where employees feel personally and professionally fulfilled. This can be achieved by offering opportunities for career development, recognizing and rewarding accomplishments, and providing meaningful work that aligns with employees' values. Satisfied employees are more likely to be engaged and motivated, which can lead to higher levels of **productivity**, **creativity**, and **innovation**. Additionally, satisfied employees are more likely to be loyal to the company, reducing turnover and the associated costs of recruiting and training new staff.

Cultivating a **growth-oriented mindset**, understanding the principles of **organizational psychology**, and prioritizing **employee well-being** are critical components in fostering an environment of innovation and productivity. By implementing strategies that promote motivation, empower leadership, encourage teamwork, and support psychological well-being, businesses can unlock the full potential of their workforce. A positive and innovative organizational culture not only enhances employee engagement but also drives the creativity and resilience necessary for long-term business success. This section will equip you with practical tools and strategies to build an organizational culture that maximizes employee potential and fosters sustainable growth for your business.

Bringing It All Together: Crafting a Scalable and Sustainable Growth Strategy

By the end of this book, you will have gained a comprehensive understanding of how to integrate **data analysis**, **mathematical models**, and **psychology** to develop a sustainable, scalable growth strategy for your business. In an increasingly complex and competitive market, this integrated approach is essential for making informed decisions that drive long-term success. Whether you're a seasoned entrepreneur or just starting out, the insights provided in these pages will empower you to navigate the challenges of today's business world.

In the chapters ahead, you've learned how to leverage **data** to identify valuable insights, track performance, and make evidence-based decisions that maximize profitability while minimizing risk. You've also explored how **mathematical models** can be applied to predict outcomes, optimize resources, and manage uncertainty—tools that allow businesses to forecast trends, allocate resources more efficiently, and plan for the future with greater confidence. More importantly, you've come to understand how **psychological principles** shape consumer behavior, influence decision-making, and can be applied to improve customer engagement, loyalty, and employee productivity.

Together, these elements form the foundation of a powerful framework for **growth strategy**—one that is both proactive and adaptable. Rather than simply reacting to market shifts or external pressures, this integrated approach enables businesses to stay ahead of the curve, continuously evolving and adjusting to meet the demands of a dynamic environment. You will now have the tools to anticipate change, identify opportunities early, and implement strategies that allow your business to **scale sustainably** and thrive in an ever-changing landscape.

As markets become increasingly complex, competition grows fiercer, and consumer expectations evolve rapidly, businesses that embrace this **holistic, data-driven approach** will be better positioned to succeed. The key lies in using **data** not just as a reactive tool but as a proactive resource for spotting trends and informing strategic decisions. By applying **mathematical models**, you'll be able to quantify risks, forecast future needs, and optimize performance in a way that drives efficiency and profitability. And by embedding **psychology** into your strategy, you'll build stronger, more meaningful connections with both customers and employees, fostering a culture of innovation and resilience.

The power of this integrated approach lies in its ability to connect the dots between diverse business functions—**marketing**, **sales**, **operations**, **customer service**, and more—into a cohesive whole. This approach ensures that your business is aligned across all levels and that every decision, from product development to customer engagement, is backed by a deep understanding of data, mathematical insights, and human behavior. By weaving these elements together, you'll be able to craft a growth strategy that is not only effective but also scalable and adaptable over time.

Most importantly, you will have developed the mindset and the tools needed to build a business that is **future-ready**—one that can anticipate change, respond proactively, and stay competitive even in the face of disruption. Whether you are managing a startup or overseeing a well-established enterprise, this framework will equip you to make informed, strategic decisions that support long-term growth and success.

The businesses that excel in the modern world are those that harness the power of **data**, **mathematics**, and **psychology** to create strategies that are both responsive and forward-thinking. By combining these three pillars, you will be empowered to build a business that can not only **survive** but **thrive** in the face of rapid change, intense competition, and evolving consumer demands. This book has given you the knowledge and tools to craft a growth strategy that is truly scalable, sustainable, and future-ready—one that will serve as a strong foundation for long-term success in the ever-changing business world.

Chapter 3: The Power of Data-Driven Decision Making

In today's competitive business environment, success is no longer determined solely by intuition, experience, or luck. Rather, it's increasingly driven by the power of data and the ability to use it effectively to inform decision-making. This chapter introduces the concept of data-driven decision making and explores how businesses can harness the power of data to gain valuable insights, optimize strategies, and drive predictable and scalable growth.

Introduction to Data Analysis in Business

Data analysis has become an indispensable tool for businesses aiming to thrive in the modern marketplace. While intuition and experience still play important roles, the sheer volume of available data—coupled with advances in technology—has made it possible for businesses to make more informed and strategic decisions. Data analysis allows companies to go beyond guesswork, offering a more precise and objective way to understand their market, their customers, and their operations. By examining past performance and current trends, data enables businesses to forecast future outcomes with greater accuracy and confidence.

At its core, data analysis involves collecting, processing, and interpreting data to extract meaningful patterns and insights. This process helps businesses understand what's working, what's not, and where adjustments need to be made. Whether it's evaluating customer behavior, assessing operational efficiency, or monitoring market trends, data analysis gives organizations the power to make decisions that are based on facts rather than assumptions or hunches.

Types of Data Businesses Can Leverage

Businesses have access to a vast array of data that can be leveraged to drive growth. The types of data that are most valuable depend on the specific goals and needs of the business, but they generally fall into three broad categories: customer behavior, financial performance, and market trends.

1. **Customer Behavior Data:** Understanding how customers interact with your business is crucial to developing effective marketing and sales strategies. Customer behavior data includes information about purchasing habits, website interactions, product preferences, and customer feedback. By analyzing this data, businesses can segment customers into different groups based on their preferences, demographics, and behaviors, allowing for more personalized marketing efforts and improved customer targeting.

   For example, tracking the path a customer takes on your website can provide insights into where they drop off or where they convert. This data can inform website optimization, marketing funnels, and product recommendations, ultimately leading to better engagement and higher conversion rates.

2. **Financial Performance Data:** Financial data is another critical source of insight. This includes metrics such as revenue, expenses, profit margins, cash flow, and overall financial health. By examining these financial indicators, businesses can assess the effectiveness of their pricing strategies, the profitability of various products or services, and the overall financial viability of the business. Financial performance data can also help businesses make more informed decisions regarding investments, budgeting, and cost management.

For instance, analyzing the cost of goods sold (COGS) against revenue can reveal opportunities for reducing expenses without sacrificing quality or customer satisfaction.

3. **Market Trends Data:** Market trends data helps businesses understand the broader landscape in which they operate. This includes data on industry developments, competitive activity, technological advancements, and changes in consumer behavior at a macro level. By analyzing market trends, businesses can identify emerging opportunities and potential threats, allowing them to stay ahead of the competition. This type of data is invaluable for strategic planning, product development, and market positioning.

   For example, if a business tracks the increasing demand for sustainable products, it can adjust its product offerings or marketing strategies to align with this trend, ensuring they remain relevant in a shifting marketplace.

Key Metrics for Measuring Business Performance

To make data-driven decisions, businesses must first identify the right metrics that provide a clear and actionable picture of performance. Metrics are essential for understanding whether a business is meeting its objectives and where improvements can be made. Here are some key metrics commonly used to measure business performance:

1. **Return on Investment (ROI):** ROI is a fundamental metric that helps businesses evaluate the profitability of investments and initiatives. It measures the return generated relative to the cost of an investment and is calculated as:

$$\text{ROI} = \frac{\text{Net Profit}}{\text{Investment Cost}} \text{x100}$$

   By tracking ROI, businesses can identify which investments are generating the most value and adjust their strategies accordingly.

2. **Customer Acquisition Cost (CAC):** CAC measures the cost of acquiring a new customer, including marketing, sales, and advertising expenses. It is calculated by dividing total acquisition costs by the number of new customers acquired during a given

period. Keeping track of CAC helps businesses understand how efficiently they are acquiring customers and whether their marketing and sales strategies are cost-effective.

$$\text{CAC} = \frac{\text{Total Sales and Marketing Cost}}{\text{Number of New Customers Acquired}}$$

3. **Customer Lifetime Value (CLV):** CLV is a metric that estimates the total revenue a customer will generate throughout their relationship with a business. It is calculated by multiplying the average purchase value, the number of purchases per year, and the average customer lifespan. CLV helps businesses understand the long-term value of acquiring and retaining customers, making it an essential metric for customer retention strategies.

$$\text{CLV} = \text{Average Purchase Value X Number of Purchases Per Year xAverage Customer lifeSpan}$$

By comparing CLV with CAC, businesses can assess the profitability of their customer acquisition strategies and optimize their efforts to maximize long-term value.

4. **Churn Rate:** Churn rate measures the percentage of customers who stop doing business with a company over a specific period. A high churn rate can indicate dissatisfaction, while a low churn rate suggests strong customer retention. Monitoring churn rate helps businesses understand how well they are meeting customer needs and whether they need to make changes to improve loyalty.

$$\text{Churn Rate} = \frac{\text{Number of Customers Lost}}{\text{Total number of Customer at the start of the period}} \text{x}100$$

5. **Conversion Rate:** Conversion rate measures the percentage of visitors to your website or store who take a desired action, such as making a purchase, signing up for a newsletter, or downloading a resource. By tracking conversion rates, businesses can identify areas of their sales funnel that need improvement and refine their marketing strategies to improve lead generation and sales.

$$Conversion\ Rate = \frac{Number\ of\ Conversion}{Total\ number\ of\ visitors} x100$$

How Data-Driven Decisions Can Lead to More Predictable and Scalable Growth

Data-driven decision making empowers businesses to move beyond reactive, instinct-based strategies. By continuously analyzing data, companies can identify trends, patterns, and potential risks before they become problems, allowing for more proactive decision-making. For example, data-driven businesses can adjust marketing campaigns based on real-time performance metrics, optimize pricing strategies by understanding consumer willingness to pay, and streamline operations by identifying inefficiencies in the supply chain.

Furthermore, by leveraging predictive models and historical data, businesses can anticipate future demand, ensuring that they have the resources and strategies in place to meet that demand when it arises. This proactive approach helps companies avoid overextension or underperformance, leading to more predictable growth.

As businesses scale, data-driven decision making becomes even more crucial. With an increasing volume of data and complexity, relying on intuition alone becomes less effective. Data analytics helps businesses maintain control and clarity as they grow, ensuring that each decision made is informed by solid evidence and contributes to scalable, long-term success.

In this chapter, we've explored the power of data-driven decision making and how businesses can leverage data to enhance their decision-making processes. By utilizing data from customer behavior, financial performance, and market trends, businesses can identify key metrics that drive performance and ultimately optimize their growth strategies. Data-driven decisions enable businesses to operate more predictably, scale more effectively, and reduce risk, ensuring long-term success in an increasingly competitive marketplace. In the chapters to come, we will dive deeper into how to build predictive models, apply psychological principles, and integrate data analysis into every aspect of your business strategy.

Understanding Mathematical Models for Scaling Businesses

Mathematical models are essential tools that enable businesses to predict, optimize, and scale their operations. By utilizing various mathematical approaches, businesses can model growth patterns, project future performance, and make more informed decisions regarding resource allocation, sales forecasting, and market penetration. This chapter introduces you to the basics of mathematical models and how they can be applied to practical business scenarios to drive sustainable growth.

Basics of Mathematical Models

Mathematical models are abstract representations that use mathematical language to describe relationships between variables. In business, these models help to simplify and quantify complex processes, allowing decision-makers to better understand trends, predict outcomes, and make data-driven decisions. While there are many types of mathematical models, some of the most relevant for business scaling are **linear growth**, **exponential growth**, and **logarithmic growth** models.

1. **Linear Growth Model**:

   A linear growth model assumes that a business grows at a constant rate over time. In this model, the change in output (e.g., sales, customers, revenue) is proportional to the time or effort invested. The formula for linear growth is:

$$y = mx + by$$

   Where:

   - y is the dependent variable (e.g., sales or revenue),
   - m is the slope or rate of change (the constant growth rate),
   - x is the independent variable (e.g., time or resources),
   - b is the starting point or initial value.

Linear growth is ideal for predicting steady, predictable increases over time and is often used in situations where growth is expected to be consistent and stable, such as in traditional retail or small businesses that are operating in a mature market.

2. **Exponential Growth Model**:

Exponential growth describes a scenario where growth accelerates over time, typically in environments that allow for rapid scaling or compounding effects. In exponential growth, the rate of growth is proportional to the current value, which leads to increasingly faster growth as time progresses. The formula for exponential growth is:

$$Y = a.e^{(bt)}$$

Where:

- y is the dependent variable (e.g., sales or revenue),
- a is the initial value,
- e is the base of the natural logarithm (approximately 2.718),
- b is the growth rate,
- t is time.

This model is often used to describe businesses in high-growth sectors, such as tech startups or viral products, where the growth rate accelerates as more customers, resources, or users are acquired. It's particularly useful for businesses experiencing network effects, where the value of the product increases as more people use it.

3. **Logarithmic Growth Model**:

Logarithmic growth represents a situation where growth slows down over time. In this model, initial growth is rapid, but as the process matures, the rate of increase diminishes. The formula for logarithmic growth is:

$$Y = a + b.\text{In}(x)$$

Where:

- y is the dependent variable (e.g., sales or revenue),
- a is the starting point or baseline,
- b is a scaling factor,
- ln(x) is the natural logarithm of the independent variable.

Logarithmic growth is often seen in markets where businesses quickly capture the majority of available customers or resources, and then the growth rate slows as the remaining market becomes saturated. This model is commonly used in industries where early adoption drives growth, but once most potential customers have been reached, growth starts to level off.

How These Models Can Be Applied to Sales Forecasting, Resource Allocation, and Market Penetration

Mathematical models play a crucial role in various business functions, particularly in sales forecasting, resource allocation, and market penetration. Here's how each model can be applied in these contexts:

1. **Sales Forecasting**:

   Sales forecasting involves predicting future sales based on historical data, trends, and market conditions. Mathematical models are particularly useful in this area because they can help businesses make more accurate projections and adjust strategies in advance.

   - **Linear Growth**: If your business operates in a mature market with consistent demand (such as a grocery store), linear models are helpful for predicting steady sales over time.
   - **Exponential Growth**: If your business is a startup with a new product or service, or if you're operating in a market with viral growth potential (like a new mobile app), exponential growth models help you estimate the rapid scaling of your customer base and sales over time.

- **Logarithmic Growth**: Once a business or product reaches a certain level of market penetration, growth tends to slow down. Using a logarithmic model, you can forecast when growth may level off and start planning for strategies to sustain long-term success (e.g., product diversification, geographic expansion, etc.).

2. **Resource Allocation**:

Resource allocation refers to how a company distributes its available resources (such as time, capital, and labor) to achieve business goals. Mathematical models help optimize resource distribution by predicting how resources should be allocated to achieve the best possible outcomes.

- **Linear Growth**: For steady, predictable growth, businesses can use linear models to allocate resources at a constant rate, ensuring that production or operational processes grow at a manageable pace without overwhelming capacity.
- **Exponential Growth**: During periods of rapid growth, exponential models can help businesses anticipate the need for additional resources to keep pace with expanding demand. For example, a rapidly growing tech company might need to increase staffing, server capacity, and infrastructure in a way that reflects the accelerating growth.
- **Logarithmic Growth**: As growth slows and the market saturates, businesses can use logarithmic models to adjust resource allocation, ensuring that they maintain profitability without over committing resources to a market that has reached its peak.

3. **Market Penetration**:

Market penetration refers to the extent to which a product or service is adopted by customers in a particular market. Mathematical models can help businesses plan their strategies for entering new markets or increasing market share.

- **Linear Growth**: For mature markets or when entering markets with well-established competitors, businesses can use linear models to predict gradual increases in market share as they build their presence over time.

- **Exponential Growth**: When introducing a groundbreaking product or service, or when leveraging viral marketing, exponential growth models can help businesses predict how quickly they can capture market share and expand their customer base. A successful viral marketing campaign or network effect can lead to rapid growth in market penetration.
- **Logarithmic Growth**: After capturing a significant share of the market, growth in market penetration typically slows. Businesses can use logarithmic models to anticipate when they might need to adjust their strategy, such as focusing on retaining existing customers or exploring new segments or niches.

Case Studies of Businesses That Have Successfully Used Mathematical Models for Growth

1. **Airbnb**:
   Airbnb's growth strategy has been significantly informed by mathematical models, particularly in terms of forecasting market demand and optimizing pricing. Airbnb uses dynamic pricing algorithms based on supply and demand, local events, and competitor prices to help hosts maximize their earnings. This data-driven approach has enabled Airbnb to scale rapidly and predict price fluctuations across different cities and seasons, driving both customer acquisition and retention.
2. **Netflix**:
   Netflix has mastered the use of data analytics and mathematical models to fuel its exponential growth. Through predictive algorithms, Netflix recommends content tailored to individual preferences, ensuring higher user engagement and retention. Their use of exponential growth models helped them project future subscriber growth as they expanded globally, and their ability to use data to forecast content preferences allows them to invest in high-demand content, optimizing both customer satisfaction and profitability.
3. **Uber**:
   Uber has leveraged mathematical models in various areas, from forecasting demand to optimizing driver allocation. During periods of high demand (such as rush hour), Uber uses real-time data and predictive models to dynamically adjust pricing (surge pricing) and allocate drivers efficiently. This approach not only helps Uber scale rapidly but also

allows the company to manage resources and customer expectations effectively during peak times.

Mathematical models are invaluable tools for businesses aiming to scale and grow in a predictable, sustainable manner. By understanding and applying models such as linear, exponential, and logarithmic growth, companies can make more informed decisions related to sales forecasting, resource allocation, and market penetration. Whether you are managing a small business or a large enterprise, leveraging the power of these mathematical tools will provide you with the clarity and foresight needed to navigate growth challenges and achieve long-term success. In the next chapter, we will explore how psychological principles can complement mathematical models, enhancing customer engagement and loyalty for even greater growth.

Chapter 4: The Psychology of Consumer Behavior

In order to understand the intricacies of consumer decision-making, it is crucial to explore the psychological factors that shape and influence these choices. Consumers do not make decisions based solely on product features or price; instead, their decisions are deeply impacted by psychological principles that affect how they perceive, process, and respond to marketing messages. Understanding these principles allows businesses to tap into the subconscious drivers of behavior, craft more effective marketing strategies, and foster stronger connections with customers. This chapter delves into how psychological principles influence consumer decision-making, the role of perception, motivation, and emotional triggers in purchasing behavior, and how businesses can use these insights to build brand loyalty and create effective marketing campaigns.

How Psychological Principles Influence Consumer Decision-Making

Consumer decision-making is a multi-faceted process influenced by a variety of psychological factors, including perception, cognition, emotions, and social influences. While many consumers believe they make decisions rationally, research shows that a significant portion of their choices is guided by subconscious feelings and biases. For instance, consumers are often influenced by heuristics, or mental shortcuts, that simplify decision-making by relying on previous experiences or general rules of thumb. These heuristics can significantly affect purchasing decisions. A classic example is the **anchoring effect**, where consumers are heavily influenced by the first piece of information they encounter, such as an initial price shown for a product, which can frame their subsequent evaluations of that product's value.

Another psychological principle that shapes decision-making is **cognitive dissonance**, which occurs when a consumer experiences discomfort after making a purchase that doesn't align with their expectations or beliefs. To reduce this discomfort, consumers may justify their decisions by focusing on the positive aspects of the product or by rationalizing the decision. This principle plays a significant role in post-purchase behavior, where customers may seek affirmation of their purchase decisions from others or even the brand itself. This explains why many businesses focus on reinforcing the satisfaction of their customers after a sale, through follow-up emails, reviews, and testimonials, which can mitigate any negative feelings and increase the likelihood of repeat purchases.

The Role of Perception, Motivation, and Emotional Triggers in Purchasing Behavior

Perception is a fundamental psychological process that affects how consumers view and interpret information. In the context of consumer behavior, perception refers to how a product, service, or brand is viewed in the consumer's mind, based on prior experiences, beliefs, and external influences. Perception goes beyond the actual attributes of a product; it involves how a consumer interprets those attributes and how they are shaped by factors like branding, packaging, and marketing communication. For instance, a consumer might perceive a product as high-quality due to its attractive packaging or premium pricing, even if the actual quality may not be superior to that of a less expensive alternative. Businesses that understand the power of perception can strategically design their offerings to enhance positive associations, create trust, and differentiate themselves from competitors.

Motivation plays a crucial role in guiding consumer behavior, as it drives individuals to act in order to fulfill their needs and desires. Motivation can be broken down into **intrinsic** and **extrinsic** factors. Intrinsic motivation refers to the internal desires to fulfill personal goals, such as self-improvement or achieving a sense of accomplishment. For example, a consumer may be motivated to purchase a gym membership to achieve better health or fitness. On the other hand,

extrinsic motivation comes from external rewards or recognition, such as receiving a discount or winning a prize. Marketers often leverage both intrinsic and extrinsic motivations by positioning products as means to meet both personal goals and provide rewards, thus increasing their appeal.

Consumers are also deeply influenced by **emotional triggers**, which are powerful psychological stimuli that can prompt immediate purchasing decisions. Marketing that appeals to emotions often drives stronger consumer responses than purely rational appeals. For example, emotions like fear, happiness, nostalgia, or excitement can push consumers to take action in ways that logic alone cannot achieve. Brands that evoke positive emotions, such as joy, belonging, or empowerment, can foster a stronger bond with their audience, creating deeper, long-lasting loyalty. Think of campaigns by brands like Coca-Cola or Nike, which associate their products with feelings of joy, motivation, or social connection. By connecting emotionally, these brands become more than just a product; they become a symbol of an experience or feeling the consumer craves.

Using Psychology to Build Brand Loyalty and Create Effective Marketing Campaigns

Once businesses understand the psychological factors driving consumer behavior, they can apply this knowledge to build brand loyalty and design marketing campaigns that resonate with their target audience. One key psychological driver of brand loyalty is **trust**, which can be established through consistent, positive brand experiences. Consumers are more likely to remain loyal to brands that they trust, and trust is often built on factors like product quality, transparency, and the alignment of brand values with those of the consumer. For instance, a company that demonstrates a commitment to sustainability may attract loyal customers who prioritize environmental responsibility. Additionally, consistent and reliable customer service is another important factor in building trust and loyalty.

Another psychological principle that plays a role in loyalty is **reciprocity**, the concept that people tend to return favors or gestures of goodwill. In business, this principle can be harnessed through reward systems, loyalty programs, and exclusive offers. When a customer receives a discount, a personalized offer, or an added benefit, they often feel compelled to reciprocate by continuing their patronage. Programs like frequent flyer miles or points systems are excellent examples of how reciprocity can drive repeated purchases and reinforce brand loyalty. By making customers feel valued and appreciated, businesses can create a cycle of positive reinforcement that leads to long-term loyalty.

Furthermore, effective **marketing campaigns** often use psychological insights to create messages and strategies that resonate with the core motivations of their target audience. Marketers can tap into universal human desires such as belonging, achievement, or security, and use these emotions to frame their products or services as solutions to those desires. For example, a campaign that highlights how a product can help someone feel more confident or successful can trigger the consumer's desire for social validation or self-improvement. Similarly, campaigns that focus on helping consumers solve problems or overcome challenges (e.g., a productivity tool or a fitness product) can create strong emotional connections with the brand.

Psychological principles such as **scarcity** and **social proof** are also powerful marketing tools. Scarcity refers to the idea that consumers place more value on products that are perceived as scarce or in limited supply. For instance, limited-time offers or exclusive products can drive consumers to act quickly due to the fear of missing out (FOMO). Social proof, on the other hand, leverages the influence of others to guide purchasing decisions. This can include testimonials, reviews, or influencer endorsements, where the behavior or opinions of others validate the consumer's choice and reduce uncertainty.

Understanding the psychology of consumer behavior is critical for businesses looking to develop strategies that drive growth and build long-term customer loyalty. By leveraging insights into perception, motivation, emotional triggers, and other psychological factors, businesses can create compelling marketing campaigns, foster meaningful connections with customers, and ultimately influence purchasing decisions. In the next chapter, we will explore how behavioral economics can further enhance the effectiveness of marketing and business strategies, providing additional tools for companies seeking to understand and influence consumer choices.

Chapter 5: Leveraging Behavioral Economics for Business Growth

In today's highly competitive and rapidly evolving business environment, merely understanding the psychological factors that drive consumer behavior is not enough. While recognizing the motivations, desires, and emotions that influence customer decisions is critical, businesses must also understand the deeper and more complex nature of **irrational behavior** that shapes how people make purchasing decisions and interact with brands. Traditional economic theory assumes that consumers are rational actors, making decisions based on a thorough analysis of available information and seeking to maximize their utility. However, **behavioral economics** challenges this rational model, offering a much more nuanced understanding of decision-making by demonstrating that people's choices are frequently influenced by cognitive biases, emotions, social pressures, and a host of other irrational factors. These biases often lead consumers to make decisions that deviate from what would be considered rational or optimal, creating both opportunities and challenges for businesses.

Behavioral economics has become an essential tool for businesses because it enables them to predict and influence consumer behavior in ways that traditional economic theory cannot. This chapter will explore key concepts within behavioral economics—such as **framing effects**, **anchoring**, and **loss aversion**—and examine how these principles can be applied to improve customer experiences, enhance sales performance, and drive overall business growth. By understanding and leveraging these irrational behaviors, businesses can move beyond conventional marketing strategies and create more effective, engaging, and persuasive tactics that resonate with customers on a deeper level.

Understanding irrational behavior is crucial because it offers businesses the opportunity to align their strategies with the real-world ways that people make decisions. Instead of relying on assumptions about how consumers should act, companies can develop marketing, sales, and pricing strategies that better reflect how people actually behave. By capitalizing on insights from behavioral economics, businesses can improve their customer engagement, increase conversion rates, and ultimately foster long-term customer loyalty. Furthermore, these principles can be used to **optimize pricing strategies**, enhance promotional tactics, and improve customer service, all of which contribute to sustainable growth and success in a competitive market.

In the following sections of this chapter, we will dive deeper into the fundamental principles of **behavioral economics** and explore how businesses can effectively apply these insights to drive growth, improve customer satisfaction, and ultimately outpace their competitors in an increasingly crowded marketplace. By leveraging the power of behavioral economics, companies can not only anticipate consumer behavior but also shape it in ways that benefit both the consumer and the business.

Key Concepts in Behavioral Economics

One of the fundamental ideas in **behavioral economics** is that **human behavior is not entirely rational** and is often influenced by psychological factors that distort judgment. Some of the most well-known concepts in this field include **framing effects, anchoring**, and **loss aversion**—all of which can have profound impacts on how consumers perceive products, prices, and choices.

1. **Framing Effects**: Framing refers to the way in which information is presented and how it can influence decision-making. People tend to respond differently to the same information depending on how it is framed. For instance, consumers are more likely to choose a product when the price is presented as a discount rather than a surcharge, even if the monetary value is the same. The psychological principle behind this is that individuals are more sensitive to potential losses than to equivalent gains, a phenomenon known as **loss aversion** (which we'll discuss in more detail shortly). The way a decision or offer is framed can dramatically alter its appeal, which is why businesses often highlight the positive aspects of a deal or present offers as limited-time opportunities to create a sense of urgency and potential loss.
2. **Anchoring**: Anchoring occurs when consumers rely too heavily on an initial piece of information (the "anchor") when making decisions. This bias can influence a variety of purchasing decisions, especially in relation to pricing. For example, if a consumer is presented with a high-priced item first, they may perceive a subsequent, lower-priced item as more affordable, even if the lower price is still relatively high. In a retail setting, anchoring is often seen in **sales** and **discounts**, where the original price (the anchor) is shown alongside a discounted price, leading consumers to believe they are getting a better deal than they might actually be. By understanding how anchoring works,

businesses can price their products in a way that maximizes perceived value and encourages purchasing behavior.

3. **Loss Aversion**: **Loss aversion** refers to the psychological tendency for people to prefer avoiding losses rather than acquiring equivalent gains. In other words, the pain of losing something is psychologically more impactful than the pleasure of gaining something of equal value. For example, consumers may be more motivated to purchase a product if they believe it will help them avoid losing out on an opportunity or benefit. This principle is often utilized in marketing and sales strategies, where businesses emphasize what the consumer stands to lose if they do not take advantage of an offer, rather than what they will gain. **Limited-time offers** or the introduction of scarcity (e.g., "Only 3 left in stock!") exploit loss aversion by making consumers feel the urgency of avoiding a loss.

How Understanding Irrational Behavior Can Help Businesses Improve Customer Experience and Sales

The beauty of behavioral economics lies in its ability to **uncover the hidden, irrational drivers** of consumer behavior, allowing businesses to improve both the customer experience and sales performance. By recognizing and incorporating these irrational factors into decision-making processes, companies can optimize their offerings to better meet consumer needs and desires.

For example, customers often make purchasing decisions based on **emotion** rather than reason, and businesses can tap into this by designing experiences that cater to emotional triggers, such as excitement, fear, or trust. **Retailers and service providers** that recognize how emotions influence behavior can create personalized and emotionally resonant customer experiences, encouraging more frequent engagement and higher sales volumes. For instance, offering an unexpected **reward or surprise discount** to loyal customers taps into the concept of reciprocity, where the customer feels compelled to return the favor by purchasing more.

Additionally, **understanding the irrational nature of decision-making** can help businesses create more intuitive and user-friendly experiences. For instance, when consumers are overwhelmed by too many choices, they tend to make poorer decisions, often resulting in decision fatigue or the abandonment of the purchase process. This is where **choice architecture** comes into play—a concept in behavioral economics that suggests businesses can simplify decision-making by narrowing the choices presented to consumers. By offering a smaller, curated selection of options, businesses can reduce cognitive overload and increase the likelihood of a purchase.

Moreover, businesses can use the insights from behavioral economics to anticipate **common cognitive biases** and adjust their sales strategies accordingly. For example, the **endowment effect** suggests that people place higher value on items they own, so allowing customers to interact with or "try out" a product before buying it can increase the likelihood of a purchase. This is commonly seen in industries like tech (where consumers may want to test out a new gadget) or fashion (where customers try on clothes). By incorporating features that allow consumers to feel as though they "own" a product temporarily, businesses can tap into the endowment effect and create stronger emotional ties to their products, increasing sales.

Real-World Applications of Behavioral Economics in Product Pricing, Promotions, and Customer Service

Now that we've established the key concepts, let's look at how these insights are applied in the real world, particularly in the areas of product pricing, promotions, and customer service.

1. **Product Pricing**: Behavioral economics has had a profound impact on how businesses price their products. Companies often use **price anchoring** to create the illusion of value. For example, offering a **premium version** of a product at a significantly higher price can make a regular version seem like a better deal. This pricing tactic is often used in subscription-based services, where consumers are shown multiple pricing tiers to encourage them to opt for a mid-tier or higher subscription plan. Additionally, businesses frequently use **psychological pricing strategies**, such as pricing products at $9.99 instead of $10, to leverage the **left-digit effect**, where consumers perceive the former as significantly cheaper than the latter, even though the difference is minimal.
2. **Promotions**: Promotions and discounts are powerful tools for businesses to encourage purchases, especially when paired with the concept of **loss aversion**. A **limited-time offer** or **flash sale** can trigger a sense of urgency, prompting consumers to act quickly to

avoid losing the opportunity. Additionally, the concept of **bundling** can tap into consumers' desire for more value. By offering a bundle at a slightly reduced price compared to purchasing items separately, businesses can use **anchoring** to make the bundle seem like a better deal, thereby increasing sales volume. Furthermore, **free trials** or **money-back guarantees** play into loss aversion by allowing consumers to experience the product or service risk-free, reducing their fear of loss while increasing the likelihood of a commitment.

3. **Customer Service**: Behavioral economics also plays a crucial role in shaping customer service interactions. **Framing customer complaints** or refund policies in a way that minimizes the perception of loss can enhance the overall customer experience. For example, offering a **refund** instead of a **store credit** can reduce the feeling of loss and make a customer feel more satisfied, even if they ultimately do not receive the product they wanted. In service-based industries, where customers' perceptions of fairness and satisfaction are critical, applying behavioral economics can improve both customer retention and overall satisfaction.

By applying the principles of **behavioral economics**, businesses can tap into the **irrational elements of consumer behavior** that often dictate purchasing decisions. Concepts like framing effects, anchoring, and loss aversion can be used strategically to optimize product pricing, enhance promotional offers, and improve customer service experiences. Recognizing that consumers do not always act rationally opens up a new realm of possibilities for businesses to better align their strategies with how people actually make decisions, resulting in increased sales, customer loyalty, and overall business growth. In the next chapter, we will explore the power of **psychological pricing** and how businesses can use it to further influence purchasing behavior and boost profitability.

Chapter 6: The Role of Machine Learning and AI in Business Scaling

In the rapidly evolving and increasingly digital marketplace, businesses must be able to leverage technology not only to stay competitive but also to **scale** effectively and **innovate** constantly. The rise of **machine learning (ML)** and **artificial intelligence (AI)** represents a game-changing opportunity for companies of all sizes to revolutionize how they operate, make decisions, and engage with customers. These technologies are no longer futuristic concepts but essential tools that businesses can use to **transform data into actionable insights**, **enhance operational efficiencies**, and **optimize customer interactions**. AI and ML are at the forefront of this transformation, helping businesses navigate an increasingly complex landscape with greater precision, agility, and insight.

Machine learning and **AI** are closely intertwined but distinct technologies. Machine learning refers to a subset of AI, involving algorithms and statistical models that enable systems to learn from data, improving their performance over time without explicit programming. AI, however, encompasses a broader field, including not only machine learning but also technologies like **natural language processing**, **computer vision**, and **robotics**, which enable machines to perform tasks that typically require human intelligence. The synergy between these technologies creates powerful tools that can analyze large volumes of data quickly, spot patterns, make predictions, and even automate decision-making processes. By applying these capabilities, businesses can **optimize marketing**, enhance **customer targeting**, improve **operational efficiency**, and innovate in ways that were previously unimaginable.

How Predictive Analytics and AI Enhance Business Operations

One of the most significant benefits of AI and machine learning in business is the ability to harness the power of **predictive analytics**. Predictive analytics involves using historical data and machine learning models to forecast future trends, behaviors, and events. For businesses, this means being able to anticipate demand, **forecast sales**, optimize inventory levels, and make better-informed decisions about staffing and production. Predictive analytics allows businesses to shift from reactive decision-making to proactive strategies, helping them stay ahead of competitors and reduce the risks associated with uncertainty.

For example, predictive analytics can help businesses anticipate **customer churn** by analyzing customer behavior patterns. By identifying warning signs of customers who are likely to leave, businesses can intervene with personalized offers or retention strategies to keep those customers. Similarly, companies can use predictive models to forecast **market trends**, enabling them to adjust their product offerings or marketing campaigns ahead of time, ensuring they meet customer needs before the demand spike occurs.

AI can also be used to create personalized experiences for customers based on predictive insights. By analyzing individual customer behaviors, preferences, and past interactions, AI can make real-time predictions about what products or services a customer is most likely to be interested in. This enables businesses to **personalize marketing campaigns**, **product recommendations**, and **customer interactions**—all of which can enhance customer satisfaction and loyalty. For instance, **e-commerce platforms** use AI-driven recommendation engines to suggest products based on a shopper's browsing history, previous purchases, or even what other customers have bought. This creates a more **engaging customer experience** and increases conversion rates.

AI-Driven Customer Targeting and Personalization

AI has also revolutionized **customer targeting**, enabling businesses to reach their ideal audiences more effectively. Machine learning algorithms analyze large datasets to identify consumer segments based on behavior, demographics, and purchasing history. This information allows businesses to create highly targeted marketing campaigns that speak directly to the needs, interests, and preferences of specific customer groups.

Beyond targeting, AI allows businesses to **personalize offerings** in real-time. **Personalization** is one of the most powerful ways to engage customers in today's digital world. With AI, businesses can tailor content, product recommendations, promotions, and even prices based on individual customer data, delivering highly relevant experiences that make customers feel valued. For example, **streaming services** like **Spotify** or **Netflix** use AI to recommend content based on a user's listening or viewing history, ensuring that every user feels like the service is tailored specifically for them. This level of personalization drives higher engagement, increases customer retention, and boosts overall satisfaction.

In retail, AI-powered solutions help brands offer **personalized shopping experiences** by providing customized product recommendations, offering promotions based on purchase history, or even suggesting additional items to complete an outfit. This personalized approach not only enhances the customer experience but also improves the company's bottom line by increasing sales and customer loyalty. Similarly, in industries like **finance** and **insurance**, AI can tailor product offerings, such as loans, credit cards, or policies, to meet the unique needs of individual customers, creating more value for the customer while increasing business opportunities.

Operational Efficiency and Automation through AI

AI and machine learning also play a crucial role in **optimizing operations** and driving efficiency across business functions. Whether in manufacturing, logistics, customer service, or human resources, AI can automate tasks, optimize workflows, and improve productivity. One of the most notable applications is in **supply chain management**, where AI can forecast demand fluctuations, optimize inventory levels, and help businesses avoid stockouts or overstocking, reducing costs and improving customer satisfaction.

AI has also made a significant impact on **customer service** through the development of intelligent **chatbots** and **virtual assistants**. These AI-powered tools are capable of handling a wide range of customer queries, from simple FAQs to more complex requests, providing instant responses and freeing up human agents to focus on more intricate issues. For instance, many businesses use AI-driven chatbots to assist customers with basic inquiries, process transactions, and even handle appointments, improving the overall **customer experience** while reducing operational costs. Moreover, chatbots can operate around the clock, providing businesses with the flexibility to engage with customers at any time, increasing customer satisfaction.

AI is also helping businesses **optimize pricing strategies**. Machine learning algorithms can analyze vast amounts of market data in real time to determine the best price points for products or services, taking into account competitor prices, market conditions, customer demand, and other factors. This dynamic pricing strategy allows businesses to **maximize revenue** while remaining competitive in the marketplace.

Examples of Companies Using AI to Scale and Innovate

Many forward-thinking companies have successfully integrated AI and machine learning into their operations, achieving impressive scaling and innovation. **Amazon** is perhaps the most iconic example of a company leveraging AI for growth. The company uses AI in a variety of ways, from **inventory management** and **predictive analytics** to personalizing customer recommendations and optimizing its logistics. Amazon's AI-powered recommendation engine is one of the most effective tools in driving sales by suggesting products tailored to individual shoppers. Additionally, Amazon's **Alexa** has redefined the smart home market, demonstrating how AI can create entirely new consumer experiences.

Another notable example is **Netflix**, which uses machine learning algorithms to personalize content recommendations and optimize its video streaming service. By analyzing viewing habits, Netflix's AI can predict what shows or movies users are likely to enjoy, which has contributed to the company's **huge subscriber growth**. Additionally, Netflix uses AI to make decisions about which content to produce based on data-driven insights into viewer preferences, allowing the company to create content that resonates with a wide audience.

In the **healthcare** sector, companies like **IBM Watson Health** have utilized AI to assist in diagnosing diseases and recommending treatment options based on an analysis of patient data and medical research. AI has the potential to revolutionize the way healthcare providers approach patient care, improving accuracy and efficiency while driving innovation in treatments and drug development.

The **financial services** industry has also seen transformative applications of AI. **JPMorgan Chase** uses AI to analyze financial data and assess risk, helping the company make more informed investment decisions. **Ant Financial**, an affiliate of **Alibaba**, uses AI for fraud detection, credit scoring, and risk assessment, offering a more personalized financial service experience for its customers.

Machine learning and artificial intelligence are no longer just advanced technologies reserved for large corporations—they are now essential tools that businesses of all sizes can use to **optimize operations**, **improve customer experiences**, and **drive growth**. By utilizing **predictive analytics**, **AI-powered customer targeting**, **personalization**, and **operational automation**, businesses can create more efficient, effective, and scalable systems. Companies that successfully integrate AI and ML into their strategies will not only be able to stay ahead of competitors but will also be able to **innovate** continuously, offering more personalized, relevant, and impactful products and services. The future of business is driven by data, and the companies that embrace AI and machine learning will be best positioned to thrive in this increasingly data-driven world. In the next chapter, we will explore how businesses can build a **culture of innovation** and **adaptability**, ensuring they remain agile and responsive in an ever-changing market.

Chapter 7: Creating a Scalable Business Model: Insights from Math and Psychology

In today's rapidly evolving business landscape, companies must develop scalable business models that not only drive growth but also resonate with their target audiences. A scalable business model enables a company to expand its operations and customer base without proportionally increasing costs, thereby achieving sustainable growth. This chapter explores how integrating mathematical efficiency with psychological appeal can lead to the creation of such models. We will examine how mathematical models can be utilized to predict growth and optimize processes, while also discussing the importance of incorporating psychological insights to attract and retain customers. By the end of this chapter, readers will understand how to design business models that are both mathematically sound and psychologically engaging, paving the way for sustained success in a competitive market.

Designing business models that are both mathematically efficient and psychologically appealing.

Designing business models that are both mathematically efficient and psychologically appealing is essential for achieving sustainable growth and maintaining a competitive edge in today's dynamic market environment. Such models not only optimize operational processes but also resonate with consumer behaviors and preferences.

Mathematical Efficiency in Business Models

Mathematical efficiency involves leveraging quantitative techniques to enhance various aspects of business operations. By employing mathematical models, businesses can:

- **Forecast Demand and Supply:** Utilize predictive analytics to anticipate market trends, enabling proactive adjustments in inventory and resource management.
- **Optimize Pricing Strategies:** Apply pricing models that balance profitability with customer willingness to pay, ensuring competitive advantage and market penetration.
- **Streamline Operations:** Implement process optimization algorithms to identify inefficiencies, reduce costs, and improve service delivery.
- **Assess Financial Risks:** Use financial modeling to evaluate investment opportunities, forecast returns, and mitigate potential risks.

These mathematical approaches facilitate data-driven decision-making, leading to improved efficiency and effectiveness in business operations.

Psychological Appeal in Business Models

Psychological appeal focuses on understanding and influencing consumer behavior to foster engagement and loyalty. Incorporating psychological insights into business models can be achieved through:

- **Personalization:** Tailoring products, services, and marketing messages to individual preferences enhances relevance and emotional connection with consumers.
- **Emotional Branding:** Crafting brand narratives and visuals that evoke positive emotions can strengthen brand identity and consumer attachment.
- **Social Proof:** Highlighting customer testimonials, reviews, and user-generated content leverages the influence of peer opinions to build trust and credibility.
- **Scarcity and Urgency:** Implementing limited-time offers or exclusive deals taps into consumers' fear of missing out (FOMO), prompting quicker purchasing decisions.
- **Reward Systems:** Establishing loyalty programs and incentives aligns with consumers' desire for recognition and appreciation, encouraging repeat business.

By integrating these psychological strategies, businesses can create compelling experiences that resonate with consumers, driving engagement and long-term loyalty.

Integrating Mathematical Efficiency with Psychological Appeal

The synergy between mathematical efficiency and psychological appeal enables businesses to:

- **Enhance Customer Segmentation:** Analyze purchasing behaviors and preferences to develop targeted marketing strategies that address specific needs and desires.
- **Optimize Customer Journeys:** Utilize data analytics to map customer interactions across touch points, identifying opportunities to personalize experiences and remove friction.
- **Predict Consumer Responses:** Employ behavioral modeling to forecast how consumers might react to different marketing stimuli, allowing for the design of more effective campaigns.
- **Balance Profitability with Customer Satisfaction:** Apply conjoint analysis to determine the optimal combination of product features and pricing that maximizes both company profits and consumer value perception.

This integrated approach ensures that business strategies are not only efficient but also resonate with consumers on a psychological level, leading to enhanced satisfaction and loyalty.

Case Study: Implementing an Integrated Business Model

Consider a retail company that combines mathematical efficiency with psychological appeal in its business model:

- **Data-Driven Inventory Management:** The Company uses predictive analytics to forecast demand for various products, ensuring optimal stock levels and reducing overstock situations.
- **Personalized Marketing Campaigns:** By analyzing customer purchase history and browsing behavior, the company sends tailored recommendations and offers, enhancing the relevance of marketing messages.
- **Loyalty Program with Behavioral Incentives:** The company implements a rewards system that not only offers discounts but also provides exclusive experiences, appealing to customers' desire for status and belonging.

- **User-Friendly Online Platform:** The website is designed based on user behavior analysis, ensuring easy navigation, personalized content, and quick checkout processes, aligning with consumers' preferences for convenience and efficiency.

By harmonizing mathematical efficiency with psychological appeal, this company creates a business model that is both operationally effective and deeply attuned to consumer needs, resulting in sustained growth and a strong market position.

Designing business models that are both mathematically efficient and psychologically appealing requires a deep understanding of quantitative analysis and consumer psychology. This integration enables businesses to optimize operations while delivering personalized and emotionally resonant experiences to consumers. The result is a robust business model capable of achieving sustainable growth, fostering customer loyalty, and maintaining a competitive edge in an ever-evolving market landscape.

How to use mathematical models to predict growth and optimize processes.

Integrating mathematical models to predict growth and optimize processes, alongside incorporating psychological insights to attract and retain customers, forms a dual strategy that can significantly enhance business performance. By leveraging quantitative analysis and understanding consumer behavior, businesses can achieve more accurate forecasting, efficient operations, and deeper customer engagement.

Using Mathematical Models to Predict Growth and Optimize Processes

Mathematical models serve as essential tools in forecasting and refining business operations. They enable organizations to translate complex data into actionable insights, facilitating informed decision-making. Key applications include:

- **Demand Forecasting:** Utilizing statistical methods such as time series analysis, businesses can predict future demand based on historical data. This allows for better inventory management and resource allocation.

- **Process Optimization:** Techniques like linear programming and simulation modeling help identify the most efficient ways to allocate resources and streamline operations, reducing costs and improving service delivery.
- **Financial Planning:** Financial models assess investment opportunities, forecast returns, and evaluate risks, supporting strategic budgeting and investment decisions.

By applying these mathematical approaches, businesses can enhance operational efficiency, make data-driven decisions, and strategically plan for growth.

Incorporating Psychological Insights to Attract and Retain Customers

Understanding consumer psychology is crucial for developing strategies that resonate with target audiences. Psychological insights inform marketing efforts, product development, and customer service enhancements. Effective strategies include:

- **Personalization:** Tailoring products, services, and communications to individual preferences fosters a sense of relevance and connection, increasing engagement and loyalty.
- **Emotional Branding:** Crafting brand narratives and visuals that evoke positive emotions can strengthen brand identity and consumer attachment, influencing purchasing decisions.
- **Social Proof:** Highlighting customer testimonials, reviews, and user-generated content leverages the influence of peer opinions to build trust and credibility, encouraging new customers to engage.
- **Scarcity and Urgency:** Implementing limited-time offers or exclusive deals taps into consumers' fear of missing out (FOMO), prompting quicker purchasing decisions and enhancing sales.
- **Reward Systems:** Establishing loyalty programs and incentives aligns with consumers' desire for recognition and appreciation, encouraging repeat business and long-term engagement.

Integrating these psychological tactics into business strategies can lead to more meaningful customer interactions, fostering loyalty and driving sustained growth.

Combining Mathematical and Psychological Approaches for Enhanced Business Strategies

The synergy of mathematical modeling and psychological insights enables businesses to:

- **Refine Targeting and Positioning:** Data analytics identify consumer segments, while psychological insights tailor messaging and offerings, ensuring alignment with audience desires and needs.
- **Enhance Customer Experience:** Predictive models anticipate customer needs, and personalized experiences meet those expectations, leading to increased satisfaction and retention.
- **Optimize Marketing Campaigns:** Analyzing campaign data informs adjustments, and understanding consumer behavior guides content and delivery, maximizing impact and ROI.

By harmonizing quantitative analysis with an understanding of consumer behavior, businesses can create comprehensive strategies that drive efficiency, engagement, and growth.

Chapter 8: Building and Managing a Growth-Oriented Team

Building and managing a growth-oriented team is crucial for achieving long-term success and sustainability in today's rapidly changing business landscape. As businesses face new challenges, ever-evolving technologies, and increased global competition, it has become clear that organizations can no longer rely solely on traditional management strategies. In order to stay competitive, organizations need to develop teams that can innovate, adapt to change, and collaborate seamlessly. A growth-oriented team isn't just a group of people who perform well—it is a team that thrives in dynamic environments, actively seeks continuous improvement, and supports each member in their individual and collective development. These teams are forward-thinking, proactive, and consistently strive for excellence.

At the heart of such teams lies leadership—leaders who inspire, guide, and empower their teams to reach their full potential. The role of leadership in a growth-oriented team cannot be overstated. Effective leaders are not only responsible for setting the strategic direction of the team but also for nurturing an environment that allows creativity and new ideas to flourish. They are tasked with creating a culture where adaptability is valued, where individuals feel psychologically safe enough to voice their opinions, and where collaboration is actively encouraged across all levels of the team. Leadership in a growth-oriented team requires a deep understanding of human behavior, the ability to leverage psychological insights, and a willingness to embrace data-driven decision-making. In this chapter, we will explore the critical elements of building and managing a growth-oriented team by examining three fundamental areas: the role of psychology in leadership and team dynamics, how to use data and psychological insights to motivate and lead teams, and the creation of a culture that nurtures innovation, adaptability, and collaboration. Through an understanding of these aspects, leaders can build teams that are not only high-performing but also resilient and capable of driving sustained growth in an ever-changing business environment.

This chapter begins with the exploration of **psychology's role** in leadership and team dynamics. Leadership is not just about directing; it's about understanding the psychological drivers that influence team behavior. Every individual within a team brings their own set of motivations, emotional responses, and interpersonal styles, which can significantly impact team dynamics. Leaders who are aware of these psychological factors can create environments that maximize individual and collective potential. By leveraging principles of emotional intelligence (EQ), leaders can foster an atmosphere of trust, collaboration, and empathy. A leader with high EQ is able to recognize and manage both their own emotions and the emotions of others, leading to better communication, conflict resolution, and overall team cohesion. Furthermore, understanding personality differences—such as those identified by tools like the Myers-Briggs Type Indicator (MBTI) or the Big Five personality traits—allows leaders to customize their approach to the needs of each team member. Each person brings different strengths, preferences, and work styles to the table, and recognizing these differences helps leaders align roles and responsibilities in a way that maximizes productivity and engagement. Motivational psychology also plays a crucial role in leadership, as understanding what drives team members—whether intrinsic motivations like a desire for personal growth or extrinsic motivations such as recognition and reward—enables leaders to implement more effective strategies for engagement and satisfaction.

In addition, psychological safety, a concept coined by researcher Amy Edmondson, is fundamental to fostering a growth-oriented team. Teams that feel psychologically safe are more likely to take risks, share innovative ideas, and provide constructive feedback. Leaders who prioritize creating a psychologically safe environment encourage their teams to engage in open communication without fear of judgment or reprisal. This not only boosts creativity and innovation but also allows individuals to learn from their mistakes and grow as professionals. When psychological safety is in place, team members feel that their contributions are valued, and they are empowered to take ownership of their work and ideas, which leads to a more engaged and motivated team.

The second major area discussed in this chapter is how to **motivate and lead teams using data and psychological insights**. Data-driven decision-making is increasingly important in the modern workplace, and when combined with psychological insights, it can help leaders make more informed, objective, and personalized management decisions. While traditional management approaches often relied on intuition or personal experience, today's leaders have access to an abundance of data that can shed light on performance trends, team dynamics, and potential areas for improvement. By tracking key performance indicators (KPIs), for example, leaders can gain insight into how well individual team members and the team as a whole are progressing toward goals. Data can help leaders identify patterns, such as which team members consistently exceed expectations or which areas are struggling. This information is valuable in crafting personalized development plans, providing targeted feedback, and ensuring that resources are allocated efficiently.

However, data on its own is not enough. Leaders must interpret that data through a psychological lens in order to understand what motivates their team and how to best support them. For instance, if data reveals that a team member is struggling with a particular task or project, the leader can use psychological insights to understand whether the challenge is skill-based, motivational, or perhaps related to interpersonal dynamics. With this understanding, the leader can provide more tailored support, whether through additional training, reassessing workload, or offering increased autonomy. Additionally, the use of **personalized motivation strategies** based on both data and psychological insights can help boost individual and team morale. A data-driven approach allows leaders to understand the specific needs and preferences of their team members, enabling them to implement motivation strategies that resonate with each person. Some may respond better to recognition and rewards, while others may find motivation in new challenges or opportunities for personal development.

Moreover, setting clear goals and establishing regular feedback loops based on performance data are crucial for maintaining momentum and ensuring continuous growth. By aligning team goals with individual performance, leaders can create a sense of accountability and ownership. These goals should be specific, measurable, achievable, relevant, and time-bound (SMART), and progress should be monitored regularly. Data provides an objective basis for feedback, ensuring that it is grounded in facts and not influenced by subjective biases. Recognizing and celebrating achievements, both big and small, motivates team members to continue working hard toward shared objectives. For instance, a well-structured gamification system, driven by data, can provide an enjoyable and engaging way to motivate team members, increase productivity, and promote friendly competition.

Finally, the third aspect of this chapter focuses on **creating a culture that encourages innovation, adaptability, and collaboration**. A growth-oriented team is one that is always learning, evolving, and looking for ways to improve. To foster this type of culture, leaders must encourage risk-taking and experimentation. In organizations where innovation is prioritized, employees feel empowered to try new ideas, challenge the status quo, and pursue creative solutions without fear of failure. Leaders must reinforce the idea that failure is a natural part of the process and an opportunity for growth. When employees feel that their creative efforts are valued—regardless of the outcome—they are more likely to contribute fresh ideas and engage in collaborative problem-solving.

Adaptability is equally important in today's fast-moving business environment. Teams that are flexible and responsive to change are better equipped to handle new challenges and opportunities. Leaders can foster adaptability by promoting a **growth mindset**, a concept introduced by psychologist Carol Dweck. A growth mindset encourages team members to view challenges as opportunities for development, to embrace learning, and to persist through obstacles. In an adaptable team, individuals are open to change, willing to learn new skills, and able to adjust quickly when faced with new information or shifting priorities.

Collaboration is the final pillar of a growth-oriented culture. The ability to work effectively together—across departments, roles, and skill sets—allows teams to leverage the diverse strengths of each individual. Leaders can encourage collaboration by fostering a sense of shared purpose and ensuring that team members are aligned toward common goals. Breaking down silos within an organization and promoting cross-functional collaboration helps to create an inclusive and cooperative environment where all voices are heard and valued. Leaders can also empower team members by providing them with the autonomy to make decisions, lead initiatives, and contribute to the team's strategic direction. Recognizing and celebrating collaborative efforts not only reinforces the importance of teamwork but also builds a sense of belonging and accountability.

Building and managing a growth-oriented team requires a multifaceted approach that combines psychological insights, data-driven leadership, and a culture of continuous learning and collaboration. Leaders must understand the psychological needs of their team members, use data to inform decisions and strategies, and create an environment that fosters innovation, adaptability, and collaboration. By doing so, leaders can cultivate teams that are not only high-performing but also resilient and prepared to meet the challenges of an ever-changing business landscape. The growth-oriented team is the foundation for long-term success, and with the right leadership, it can propel the organization toward its most ambitious goals.

The Role of Psychology in Leadership and Team Dynamics

Psychology plays an essential role in understanding and managing team dynamics. Leaders who recognize the psychological factors that influence team behavior can create an environment that maximizes individual and collective performance. Emotional intelligence (EQ), for example, is a key component in leadership. It enables leaders to understand their own emotions and those of their team members, facilitating better communication, trust, and conflict resolution. When leaders are emotionally aware, they can navigate sensitive situations, provide constructive feedback, and maintain a positive work environment. EQ helps to foster a sense of psychological safety, where team members feel safe to share ideas and take risks without fear of judgment. This, in turn, enhances creativity and collaboration.

In addition, understanding individual personalities through psychological tools, such as the Myers-Briggs Type Indicator (MBTI) or the Big Five personality traits, allows leaders to tailor their management style to suit the strengths, preferences, and behaviors of each team member. This personalized approach enables leaders to put individuals in positions where they can perform their best. Recognizing motivational drivers is another psychological aspect that is vital to leadership. Each person is driven by different factors—some are motivated by intrinsic factors like personal growth, while others are motivated by extrinsic factors such as recognition and rewards. Leaders who understand these differences can create strategies that align with what drives each team member, resulting in enhanced engagement and productivity.

Psychological safety is another crucial concept. A culture where team members feel comfortable speaking up, sharing ideas, and making mistakes is essential for innovation and team cohesion. In such an environment, individuals are more likely to contribute creatively, collaborate openly, and approach challenges with a growth mindset. When leaders foster psychological safety, they build a foundation of trust and openness, leading to a more cohesive, high-performing team.

How to Motivate and Lead Teams Using Data and Psychological Insights

Combining psychological insights with data-driven decision-making is an effective way to motivate and lead teams. While psychology provides an understanding of individual and group dynamics, data helps provide measurable insights that can guide leadership decisions and performance management. Data, such as key performance indicators (KPIs), can highlight strengths and areas for improvement within the team, helping leaders make informed decisions. Regular tracking and analysis of team performance allow leaders to set realistic, data-backed goals, monitor progress, and make adjustments when needed. Data also helps identify patterns of behavior that may not be immediately apparent, such as team members who consistently underperform or excel in specific areas.

Using data-driven insights, leaders can implement strategies that are more likely to resonate with their team's needs. For example, if data reveals that certain team members thrive when given autonomy, while others perform better with structured guidance, leaders can adapt their approach to ensure that each person is working in an environment that maximizes their strengths. This leads to increased motivation and engagement. Additionally, leaders can create personalized development plans based on both psychological insights and performance data, ensuring that each team member receives the support and resources they need to grow.

Another important aspect is using data for goal setting and feedback. By setting SMART (Specific, Measurable, Achievable, Relevant, Time-bound) goals, leaders can ensure clarity and alignment with the overall objectives of the team. Data helps track progress toward these goals, allowing for timely feedback and adjustments. Providing regular feedback, based on data, encourages a continuous learning process, reinforcing positive behavior and addressing areas that need improvement. Data-driven feedback ensures that the focus is on performance and progress, making it objective and actionable.

Gamification is another strategy that blends data with psychological principles. Leaders can introduce competitive elements, rewards, and recognition based on data to keep team members motivated and engaged. By making performance visible and rewarding effort and achievement, gamification taps into psychological motivators, such as competition and recognition, to boost morale and productivity. However, it's important to strike a balance and ensure that gamification remains positive and inclusive, rather than creating unhealthy competition.

Creating a Culture That Encourages Innovation, Adaptability, and Collaboration

A growth-oriented team thrives in a culture that encourages innovation, adaptability, and collaboration. To create such a culture, leaders must focus on fostering an environment where creativity is valued, risks are seen as opportunities for learning, and collaboration is prioritized. Innovation is at the heart of growth, and it flourishes in environments where team members feel empowered to experiment with new ideas and approaches. Leaders should communicate that failure is an inevitable and valuable part of the innovation process. When team members know that mistakes will not result in punitive consequences, they are more likely to take risks, explore new solutions, and push boundaries. Rewarding creativity and innovative thinking—whether through recognition or tangible rewards—reinforces this mindset and drives the team toward continued growth.

Adaptability is another key trait that defines a growth-oriented team. In a rapidly changing business environment, teams must be flexible and open to change in order to thrive. Leaders can foster adaptability by encouraging a growth mindset—one that views challenges as opportunities to learn and grow. Leaders should emphasize the importance of perseverance and resilience, helping team members understand that success comes through dedication and the willingness to learn from setbacks. Training and development opportunities are crucial for maintaining adaptability. By encouraging team members to continually develop new skills and knowledge, leaders ensure that the team remains responsive to changes in the business environment and industry.

Collaboration is equally important for creating a growth-oriented team. In today's interconnected world, success often depends on how well individuals work together across functions, departments, and even organizations. Leaders can promote collaboration by breaking down silos and encouraging cross-functional teams to work together on projects. Regular communication and joint problem-solving sessions foster an environment where diverse perspectives are heard and respected. This exchange of ideas enhances creativity, improves decision-making, and strengthens the team's ability to adapt to new challenges.

Additionally, leaders should empower team members to take ownership of their work, provide input into decision-making processes, and contribute to the overall direction of the team. Empowering individuals builds a sense of responsibility and investment in the team's success. Recognition plays a significant role in fostering a collaborative culture. By acknowledging both individual and collective achievements, leaders reinforce the value of teamwork and encourage a sense of pride and accountability in the team's shared success.

Building and managing a growth-oriented team requires a deep understanding of psychological principles, a data-driven approach to leadership, and a commitment to cultivating a culture that promotes innovation, adaptability, and collaboration. By leveraging emotional intelligence, understanding individual motivations, and promoting psychological safety, leaders can create an environment where team members feel valued, understood, and motivated to contribute their best work. Data-driven leadership further enhances the effectiveness of team management by providing objective insights into performance, helping to set clear goals and track progress, and tailoring strategies to individual needs. Lastly, fostering a culture of continuous learning, innovation, and collaboration ensures that the team remains adaptable and forward-thinking. In this way, leaders can build a team that not only achieves success but continuously grows and thrives in an ever-changing world.

Chapter 9: The Importance of Customer Feedback and Iteration

In today's customer-centric business world, feedback is one of the most valuable sources of information that organizations can tap into for continuous improvement and innovation. As businesses seek to adapt and stay competitive, understanding the importance of customer feedback and how to leverage it effectively becomes a critical component of growth and sustainability. Customer feedback serves not only as a gauge of current performance but also as a roadmap for enhancing products, services, and overall customer experiences. When paired with an iterative approach, customer feedback allows businesses to stay responsive, anticipate market shifts, and reduce the risk of stagnation. This chapter will explore how customer feedback can be used as a data source for business improvement, the psychological techniques that can improve customer satisfaction and gather deeper insights, and how continuous iteration based on this feedback can drive sustainable growth.

Using Customer Feedback as a Data Source for Business Improvement

Customer feedback provides organizations with a direct and authentic lens into the experiences and perceptions of their target audience. Whether gathered through surveys, reviews, social media comments, or customer service interactions, feedback serves as an invaluable source of data that reflects the true sentiments of customers. When businesses collect and analyze this data, they gain critical insights into what works well and what areas need improvement.

The first step in using customer feedback as a data source is ensuring that businesses have a structured system for collecting feedback. This could include a combination of qualitative and quantitative methods, such as online surveys, focus groups, interviews, or customer satisfaction ratings (e.g., Net Promoter Score or NPS). Each of these data points provides unique insights into different aspects of the customer experience. For instance, NPS scores can measure customer loyalty and likelihood of recommending a business to others, while open-ended responses can uncover nuanced opinions or concerns that may not be captured by numeric data alone.

Once feedback is gathered, it is essential to analyze it thoughtfully. Aggregating feedback from different channels allows organizations to identify common themes, recurring issues, or opportunities for improvement. By categorizing the feedback into actionable insights, businesses can pinpoint areas where they are excelling and those that may require adjustments. Moreover, feedback provides a real-time gauge of customer satisfaction, enabling businesses to react quickly to negative experiences and course-correct before dissatisfaction turns into customer churn.

Feedback also allows businesses to monitor the effectiveness of changes or new initiatives. For example, after launching a new feature or making a process improvement, businesses can track customer feedback to determine whether the changes have had the desired impact. This process of gathering feedback before, during, and after a change is essential for measuring the success of any business decision. Regularly analyzing customer feedback ensures that businesses are always aligned with customer needs, which is fundamental to long-term growth and relevance in the marketplace.

Applying Psychological Techniques to Improve Customer Satisfaction and Gather Valuable Insights

To truly unlock the potential of customer feedback, businesses must consider not only the content of the feedback but also the psychological factors that shape the way customers provide and perceive feedback. Understanding psychological principles allows organizations to tailor their strategies in ways that foster positive customer experiences and encourage valuable, actionable insights.

One key psychological principle is **reciprocity**, the idea that people are more likely to give feedback when they feel they are receiving something of value in return. When businesses show customers that their opinions matter and are being acted upon, they create a sense of mutual respect and trust. Acknowledging feedback—whether positive or negative—by responding promptly and making visible improvements based on customer input can make customers feel valued, which in turn leads to more meaningful feedback in the future. Incentivizing customers with rewards, discounts, or exclusive offers in exchange for their feedback is another effective psychological strategy that motivates customers to share their thoughts while fostering loyalty.

Another psychological factor is **cognitive dissonance**, which occurs when customers have conflicting feelings about a product or service. This often leads to ambiguous feedback, where a customer may express dissatisfaction without fully understanding the root cause of their discontent. In these cases, businesses can use techniques like follow-up questions or personalized outreach to dig deeper into the issue and uncover the underlying reasons for dissatisfaction. By actively listening and engaging with customers, businesses can reduce cognitive dissonance and collect more detailed, accurate insights that guide improvements.

Emotional intelligence (EQ) is also a critical psychological tool in improving customer satisfaction. By empathizing with customers' emotions and understanding their frustrations or needs, businesses can build rapport and create positive experiences. When customers feel that a company understands and cares about their feelings, it leads to stronger relationships and increases customer retention. For instance, when responding to negative feedback, a business should acknowledge the customer's emotions, apologize sincerely, and offer a solution. This not only resolves the immediate issue but also creates a positive emotional connection, turning a potentially negative experience into a memorable and rewarding one.

Finally, **social proof**—the psychological tendency for people to follow the actions or opinions of others—can be used strategically in gathering insights. When customers see that others are leaving feedback, particularly positive reviews, they are more likely to provide their own. Highlighting customer testimonials, user-generated content, or customer success stories on a website or social media page can encourage others to share their experiences. This creates a cycle of feedback that helps businesses continuously improve while building trust with potential customers.

How Continuous Iteration Can Fuel Sustainable Growth and Reduce the Risk of Stagnation

The concept of **iteration**—the process of making incremental improvements based on ongoing feedback—is key to ensuring that a business remains adaptable and innovative. In a world where market conditions, customer preferences, and technological advancements are constantly evolving, continuous iteration ensures that businesses stay relevant and competitive. It allows organizations to stay attuned to customer needs and rapidly respond to shifting expectations or emerging trends.

The iterative process involves making small, manageable adjustments to products, services, or processes based on feedback, analyzing the impact of those changes, and then making further improvements. This cycle of feedback, adaptation, and refinement is far more effective than waiting for large-scale changes or trying to implement a complete overhaul of a product or service. By focusing on incremental progress, businesses reduce the risk of making drastic mistakes or alienating their customer base. Iteration allows companies to experiment with new ideas and refine them based on real customer input, thus minimizing the risk of stagnation or failure.

Furthermore, continuous iteration builds a culture of innovation within the organization. It empowers teams to test new ideas, learn from mistakes, and continuously enhance their offerings. A business that iterates regularly can stay ahead of its competitors by being more agile and responsive to the market. This constant process of improvement builds customer loyalty, as customers are more likely to stay with a brand that listens to their feedback and consistently strives to meet their evolving needs.

The impact of iteration on **sustainable growth** is profound. By adopting a mindset of continuous improvement, businesses can achieve long-term success without relying on short-term fixes or one-off solutions. Customers see that their feedback is valued and integrated into the evolution of the product or service, which builds trust and enhances customer satisfaction. Moreover, continuous iteration encourages businesses to experiment with new approaches, such as diversifying their product line, testing new marketing strategies, or enhancing customer service practices, all of which contribute to growth. As a result, businesses not only meet current customer expectations but also anticipate future needs and stay ahead of market trends.

By keeping feedback loops open and focusing on continuous improvement, businesses reduce the risk of stagnation, as they are always in a state of flux, adapting to new insights and opportunities. Whether it's a small tweak to a user interface, a refinement in customer service processes, or the introduction of a new feature based on customer feedback, iteration ensures that a business never becomes complacent. In doing so, it helps to create a culture of proactive growth, ensuring that the business remains competitive, relevant, and in tune with its customers.

Customer feedback and iteration are foundational to the growth and sustainability of any business. By leveraging customer feedback as a powerful data source, applying psychological techniques to improve satisfaction and gather valuable insights, and committing to continuous iteration, businesses can create an environment of ongoing improvement. This cycle of feedback and iteration not only fuels innovation but also helps businesses stay responsive to customer needs and market demands, reducing the risk of stagnation and ensuring that growth is sustained over time. In a world where customer expectations are constantly shifting, those organizations that can listen, adapt, and evolve through iteration will continue to thrive in the competitive marketplace.

Chapter 10: Scaling Across Multiple Markets

As businesses grow and pursue opportunities beyond their initial markets, the ambition to scale across multiple markets—especially international ones—becomes a key milestone in their evolution. However, this process is not without its unique set of challenges, intricacies, and risks. Successfully scaling globally demands much more than the adoption of a straightforward business strategy. It involves an intricate blend of effective growth models, a deep understanding of market dynamics, and a keen grasp of psychological and cultural factors that significantly influence consumer behavior across various regions. While the prospect of tapping into new markets offers enormous potential for business growth, it also requires businesses to be mindful of the distinct characteristics and complexities that come with each region.

The challenges of scaling are amplified when businesses attempt to enter new international markets. These markets often present a diverse range of consumer preferences, economic conditions, cultural expectations, and regulatory environments. The companies that thrive are those that can weave together a comprehensive, multifaceted approach to scaling. This approach not only includes well-established growth models and market research, but also incorporates psychological insights and local adaptations to cater to the unique preferences and behaviors of local consumers. This chapter delves deeply into these elements, illustrating how businesses can leverage a combination of analytical frameworks and psychological strategies to scale effectively across multiple markets. By harmonizing data-driven strategies with emotional intelligence and cultural awareness, businesses can successfully navigate the complexities of scaling and achieve sustainable growth and long-term success across global markets.

Applying Growth Models and Psychological Insights to International and Multi-Market Scaling

Scaling across multiple markets, particularly international ones, requires a combination of solid growth models and an understanding of the psychological factors that drive consumer behavior. Growth models typically involve frameworks such as the **Ansoff Matrix**, **product-market fit**, and **the growth hacking model**, all of which offer strategic insights on how businesses can expand into new markets. However, when scaling globally, the psychological dynamics of each market must also be considered, making it essential to blend these models with psychological insights.

For example, growth models often emphasize the importance of **market penetration**, **market development**, **product development**, and **diversification** as strategies for scaling. While these frameworks are foundational for any business looking to scale, psychological insights help to refine these strategies. When considering market penetration, a business must understand the psychological factors influencing purchasing decisions in new markets—what drives local consumers, how they perceive foreign brands, and what emotional triggers might influence their behavior. In some regions, brand loyalty may be driven by factors such as social status, while in others, consumers may place more value on price and functionality.

Psychological insights can also aid businesses in managing the challenges of cross-cultural communication. For instance, the **Cultural Dimensions Theory** by Geert Hofstede provides insight into cultural differences across countries and can guide businesses in tailoring their marketing and operational strategies to local customs, values, and behaviors. While data-driven growth models provide the "what" and "how" of scaling, psychological insights provide the "why" behind consumer behaviors, enabling businesses to approach international expansion with a deeper understanding of local preferences and motivations.

The ability to blend these growth models with psychological insights enables businesses to create tailored marketing campaigns, products, and services that resonate with diverse customer bases across multiple regions. Rather than applying a one-size-fits-all approach, businesses can utilize local knowledge and emotional intelligence to craft market-specific strategies that meet the unique needs and desires of each target market.

The Role of Market Research, Customer Psychology, and Local Adaptations

Market research is a fundamental tool in understanding both the macro and micro factors that influence success when scaling across multiple markets. Before entering a new market, businesses must conduct comprehensive research to understand the landscape, competition, consumer behaviors, and local regulations. Market research provides a data-driven foundation for strategic decisions, but it should be complemented by an understanding of **customer psychology** to ensure that the business can meet the emotional and cultural needs of local consumers.

Customer psychology refers to the study of how consumers think, feel, and behave in relation to products, services, and brands. This is especially important when scaling internationally, as customer behaviors can vary significantly across different cultures. For instance, a marketing campaign that is successful in the United States might not have the same impact in Japan or the Middle East. In some cultures, advertisements that focus on individual success and achievement might resonate, while in others, messages emphasizing community, family, and tradition may be more effective.

Additionally, customer psychology can influence how businesses approach product pricing, customer service, and communication. In some markets, consumers may be more price-sensitive, while in others, the value of premium products or services could be more appealing.

Understanding these psychological drivers enables businesses to make informed decisions about product offerings, service models, and pricing strategies tailored to local markets.

Furthermore, **local adaptations** are essential for scaling globally. These adaptations are adjustments made to products, services, marketing strategies, and business operations to align with the unique characteristics of each market. For instance, global fast-food chains like **McDonald's** often adapt their menu offerings to cater to local tastes and dietary preferences. In India, McDonald's offers a variety of vegetarian options that align with local eating habits and cultural preferences. Similarly, companies such as **Coca-Cola** and **Nike** have localized their advertising campaigns to reflect local values and symbols, making them more relatable to customers in different regions.

Local adaptations also extend to customer service, payment methods, and distribution channels. In some regions, businesses may need to offer multilingual customer support or partner with local retailers and distributors to reach customers effectively. In regions where mobile payments are more common than credit cards, businesses may need to adjust their payment methods to align with local preferences. These local adaptations ensure that a business remains relevant and accessible to customers, building trust and loyalty in the process.

Examples of Successful Businesses Scaling Globally Using a Blend of Math and Psychology

Several businesses have successfully scaled globally by integrating both data-driven strategies and psychological insights. One notable example is **Netflix**. Initially starting as a DVD rental service in the United States, Netflix has expanded its reach to over 190 countries. The company's success in international markets can be attributed to its ability to leverage growth models such as **market development** (expanding to new regions) and **product development** (creating localized content), while also using customer psychology to craft tailored experiences for its diverse audience. Netflix's ability to recommend personalized content to viewers based on viewing history and preferences is rooted in data analytics, but its understanding of cultural differences in entertainment preferences—such as the popularity of Bollywood films in India or local dramas in Korea—demonstrates a keen awareness of local consumer psychology.

Similarly, **Airbnb** has scaled globally by blending growth models with a deep understanding of local customer psychology. Airbnb's platform, which allows homeowners to rent out their properties to travelers, used market research to understand the preferences of different customer segments in various regions. By leveraging this research, Airbnb tailored its marketing to highlight different aspects of the travel experience, whether it be affordability, luxury, or unique local experiences. Understanding that trust is a key psychological component in the decision to book accommodations, Airbnb adapted its platform with features like reviews, ratings, and detailed profiles to help build consumer confidence in new markets. In addition, the company adjusted its customer service operations to meet the specific needs of local markets, such as offering support in multiple languages and providing assistance with local regulations.

Another excellent example is **Uber**, which has successfully expanded into numerous international markets by using a mix of data-driven growth strategies and psychological insights. In many markets, Uber adapted its pricing model based on local economic conditions and consumer preferences. The company also recognized the importance of **convenience and safety**, psychological drivers that are important across cultures. By using data to optimize routes and reduce wait times, Uber enhances the convenience factor, while features like driver background checks and in-app safety measures cater to the universal psychological need for safety and security. Additionally, Uber's flexible business model—allowing local entrepreneurs to drive for the platform—has helped it scale rapidly in different regions by aligning with local economic conditions and entrepreneurial mindsets.

Scaling across multiple markets requires businesses to think strategically and adapt to the unique psychological and cultural characteristics of each region. By combining growth models with psychological insights, businesses can gain a deeper understanding of customer behavior and create tailored experiences that resonate with diverse audiences. Successful scaling across markets relies on the ability to conduct thorough market research, apply psychological principles to improve customer engagement, and make local adaptations that foster loyalty and trust. As demonstrated by companies like Netflix, Airbnb, and Uber, blending data-driven growth strategies with an understanding of customer psychology is essential for achieving global success. Through this holistic approach, businesses can scale effectively, reduce the risk of missteps, and drive sustainable growth in new and international markets.

Chapter 11: Overcoming Challenges in Business Growth

Business growth is a thrilling yet often daunting journey, with numerous obstacles that can challenge even the most well-prepared companies. As businesses scale and enter new markets or increase their operations, they frequently face complexities ranging from financial constraints, operational inefficiencies, market competition, and talent acquisition, to more psychological and emotional challenges like maintaining company culture and employee morale. Scaling requires resilience, agility, and the ability to adapt to new conditions and demands. This chapter explores the common challenges businesses face during growth and presents effective strategies for overcoming them by combining data-driven approaches and psychological insights. Additionally, it delves into the role of **risk management**, **decision-making under uncertainty**, and **mathematical models** in helping businesses navigate these challenges. Finally, we'll examine case studies of successful businesses that have overcome significant growth barriers, illustrating the practical application of these strategies in real-world contexts.

Common Challenges in Business Scaling and How to Address Them with Data and Psychological Strategies

Scaling a business is a multifaceted process that often brings several challenges. Among the most common obstacles are **financial constraints**, **operational inefficiencies**, **talent acquisition**, and **maintaining company culture** as the business grows. These challenges can be exacerbated by the psychological toll that comes with managing growth, both for business leaders and employees.

Financial Constraints: One of the most significant challenges in scaling a business is securing the necessary capital for expansion. Businesses may struggle to finance new initiatives, hire the right talent, or invest in infrastructure. To address this challenge, companies must rely on data to manage financial risks effectively and ensure they allocate resources efficiently. By applying **financial models**, businesses can predict future cash flows, assess the viability of scaling initiatives, and prioritize investments based on expected return on investment (ROI). Additionally, **psychological strategies** such as fostering an ownership mindset among employees and seeking input from key stakeholders can help ensure that financial decisions are made collaboratively and transparently.

Operational Inefficiencies: As a company grows, its operations can become more complex, and inefficiencies can arise. This could manifest in production bottlenecks, supply chain disruptions, or difficulties in maintaining consistent quality control across new markets. To address operational inefficiencies, businesses can turn to data-driven **process optimization** techniques. Utilizing **lean management principles**, **Six Sigma methodologies**, and **data analytics**, companies can identify inefficiencies, streamline processes, and maintain high levels of productivity. Additionally, leaders should focus on building a psychologically safe environment where employees feel empowered to report inefficiencies or suggest improvements without fear of reprimand. This can create a culture of continuous improvement and innovation, helping to address inefficiencies in a proactive manner.

Talent Acquisition and Retention: As businesses scale, the demand for skilled talent increases. Finding and retaining the right employees is critical, but it becomes more difficult as the business grows. Often, businesses struggle with the **cultural fit** of new employees, particularly as they expand internationally. To overcome this challenge, companies must use data to drive their **recruitment strategies**. By analyzing past hiring data, understanding the competencies that lead to success, and using predictive hiring tools, businesses can identify the best candidates more efficiently. Psychological strategies, such as emphasizing strong leadership development programs, offering competitive compensation packages, and creating employee engagement initiatives, can help with talent retention. Companies that prioritize their employees' psychological well-being through initiatives like wellness programs or flexible work environments are more likely to retain top talent as they scale.

Maintaining Company Culture: One of the more intangible but important challenges of scaling a business is maintaining its culture. As businesses grow, there is often a risk that the core values, ethos, and spirit of the company may get diluted or lost. Maintaining a strong company culture requires intentional effort, especially when businesses enter new markets or hire new teams. Leadership can use **psychological strategies** to maintain culture by embedding company values into every aspect of the business, from recruitment and onboarding to day-to-day operations. For example, leaders can use **employee engagement surveys** to gauge the company's culture and ensure employees feel connected to the company's mission. They can also invest in **leadership development programs** that ensure managers embody the company's values and are capable of fostering a cohesive work environment. By combining data-driven insights with a focus on emotional intelligence and leadership, businesses can maintain a strong culture while scaling.

Risk Management and Decision-Making Under Uncertainty Using Mathematical Models

Scaling a business comes with a variety of risks. These can include market risks (such as shifts in demand), financial risks (such as changes in interest rates or cash flow problems), operational risks (such as breakdowns in supply chains), and competitive risks (such as new entrants or price wars). Managing these risks effectively is critical to sustaining growth and ensuring the long-term success of the company.

Risk management involves the identification, assessment, and prioritization of risks, followed by the application of resources to minimize or mitigate the impact of these risks. **Mathematical models** are invaluable tools for risk management, allowing businesses to quantify the potential impact of risks and make more informed decisions. Models like **Monte Carlo simulations**, **decision trees**, and **sensitivity analysis** provide businesses with insights into the probabilities of different outcomes, enabling them to assess the best course of action under uncertainty.

For instance, a company contemplating expanding into a new international market might use **Monte Carlo simulations** to model different scenarios of market entry, considering variables like exchange rates, regulatory changes, and consumer behavior. By running these simulations, the business can determine the most likely outcomes and adjust its strategy accordingly. This quantitative approach to decision-making helps business leaders reduce uncertainty and make data-driven decisions in the face of risk.

Psychological strategies also play a key role in decision-making under uncertainty. Business leaders often face **cognitive biases**—such as overconfidence or anchoring—when making decisions under pressure. These biases can cloud judgment and lead to suboptimal decisions. By recognizing these biases, leaders can use techniques such as **premortem analysis** (imagining a scenario where a decision has failed and working backward to identify potential reasons) or **scenario planning** (preparing for a range of possible outcomes) to make more rational and less emotionally-driven decisions. Additionally, encouraging a **growth mindset** within the team can help individuals view challenges as opportunities to learn rather than threats to success.

Case Studies of Businesses That Overcame Significant Growth Barriers

Numerous companies have faced significant challenges while scaling, but their ability to overcome these barriers offers valuable lessons in resilience and strategic thinking. Let's explore a few examples of businesses that effectively navigated growth hurdles using a blend of data and psychological strategies.

1. Starbucks: One of the most notable examples of overcoming growth challenges is **Starbucks**. When the company decided to expand internationally, it faced numerous barriers including cultural differences, supply chain complexities, and competition from local coffee shops. Starbucks used a combination of market research and **psychological insights** to tailor its offerings to local tastes while maintaining its core brand identity. For instance, in China, where tea culture is deeply ingrained, Starbucks introduced a range of tea-based beverages alongside its coffee offerings, appealing to local preferences. The company also used **data analytics** to optimize store placement and product offerings based on regional demand. Additionally, Starbucks worked to maintain its **company culture** during rapid expansion by investing heavily in training programs for employees, ensuring that they felt connected to the brand's mission. This strategic combination of data-driven decisions and psychological understanding helped Starbucks overcome the challenges of scaling internationally.

2. Amazon: Amazon is another example of a company that overcame significant barriers during its expansion. Initially, Amazon was an online bookstore, but over time, it evolved into a global e-commerce and cloud computing giant. One of the challenges Amazon faced was maintaining its operational efficiency as it scaled. The company addressed this through **data-driven logistics and supply chain optimization**, implementing advanced algorithms to predict demand and optimize inventory management. Additionally, Amazon faced challenges in expanding its marketplace internationally due to different consumer behaviors and regulatory environments.

To address this, Amazon leveraged its **psychological insights** to create localized shopping experiences and ensure that its brand was perceived as both reliable and convenient across different cultures. Amazon also focused on **customer-centric innovation**, offering features like personalized recommendations, fast delivery, and a customer-first mentality that aligned with psychological needs for convenience and satisfaction.

3. Airbnb: When **Airbnb** first expanded into international markets, the company faced significant challenges related to **trust and safety**. As a platform that allows strangers to rent out their homes, trust was a major psychological barrier to adoption. Airbnb used a combination of **data analytics** and **psychological insights** to address this challenge. By implementing a robust review and rating system, Airbnb created social proof that helped build trust among users. Additionally, the company worked to ensure that its hosts felt valued and supported, offering resources like insurance coverage for property damage and a 24/7 customer support system. This commitment to safety and customer service helped Airbnb build credibility in new markets, eventually leading to its global success.

Scaling a business is a journey filled with challenges, but these challenges can be overcome by combining data-driven approaches with psychological strategies. By identifying common scaling barriers such as financial constraints, operational inefficiencies, talent acquisition, and maintaining company culture, businesses can use mathematical models and psychological insights to navigate these obstacles effectively. Risk management and decision-making under uncertainty are essential components of successful scaling, and the use of data models, along with an awareness of cognitive biases, can improve decision-making processes. Case studies from businesses like Starbucks, Amazon, and Airbnb show how strategic use of both analytical and psychological approaches can help companies surmount growth barriers and achieve sustainable, global success.

Chapter 12: Measuring Success and Continuous Improvement

Success in business growth is not merely defined by the ability to scale quickly; it is also determined by how well a company sustains and adapts to the ever-changing business landscape. Measuring success and driving continuous improvement are critical for ensuring long-term growth and resilience. To effectively assess growth, businesses must identify the right **Key Performance Indicators (KPIs)** and metrics, which act as a compass guiding organizations toward their goals. However, success is not only about tracking numerical results—it also involves fostering a **growth mindset** within the team and making adjustments based on evolving data and psychological insights. This chapter will explore the importance of defining and measuring growth success, the role of psychology in maintaining continuous improvement, and how to adapt business strategies based on a dynamic blend of data-driven insights and psychological understanding.

How to Define and Measure Growth Success Using Key Performance Indicators (KPIs) and Metrics

One of the first steps in measuring success is determining what "success" looks like for a business. Growth success isn't a one-size-fits-all concept; it varies based on the specific objectives of the organization, the stage of its development, and its market positioning. **Key Performance Indicators (KPIs)** and metrics are invaluable tools for translating abstract business goals into measurable outcomes. By identifying the right KPIs and tracking the relevant metrics, companies can assess their progress and adjust their strategies as needed.

Defining KPIs: The process of selecting KPIs begins with a clear understanding of the company's strategic objectives. These could range from **financial goals**, such as increasing revenue or profit margins, to **operational goals**, such as improving customer satisfaction or reducing churn. In the early stages of growth, businesses often focus on **customer acquisition** and **market penetration**. In later stages, KPIs may shift toward **customer retention**, **brand loyalty**, and **operational efficiency**.

Financial Metrics: These KPIs often include metrics like **revenue growth**, **profit margins**, **return on investment (ROI)**, and **cash flow**. By tracking these financial indicators, businesses can gauge the effectiveness of their scaling strategies and assess whether their growth is sustainable. Financial metrics help identify how well the company is managing its resources and where adjustments need to be made, particularly in the allocation of capital and operational costs.

Customer-Focused Metrics: As customer satisfaction is critical to sustained growth, KPIs focused on customer experience and retention play an important role in determining success. **Net Promoter Score (NPS), customer satisfaction (CSAT), customer lifetime value (CLTV)**, and **customer churn rate** are all valuable indicators of how well a company is meeting the needs of its target audience. These metrics allow businesses to understand their customers' level of loyalty, satisfaction, and overall engagement with the brand. Monitoring customer behavior and sentiment also provides businesses with actionable insights into areas for improvement.

Operational Efficiency Metrics: Scaling often brings challenges related to efficiency. Metrics such as **employee productivity**, **inventory turnover**, and **supply chain performance** can indicate how well a company is optimizing its operations. For example, if inventory turnover is low, this might indicate an overstock issue, whereas high turnover may signal supply chain inefficiencies. Similarly, by tracking **employee engagement**, businesses can assess how well their teams are managing increased workloads or adapting to new systems.

Innovation Metrics: For businesses focused on continuous improvement and staying ahead of competitors, measuring innovation is crucial. **R&D expenditure, new product development timelines**, and the percentage of revenue generated from new products are all metrics that can help gauge innovation success. These metrics reveal whether the company is investing effectively in new opportunities and whether innovation is aligned with market needs.

In addition to selecting the right KPIs, businesses must also ensure that these metrics are tracked and reviewed consistently. **Real-time dashboards** and **data analytics tools** help businesses gain a continuous overview of their performance, providing timely insights that can influence decision-making. Setting **regular review cycles** and conducting **performance audits** ensures that businesses stay on course and are able to pivot when necessary.

The Role of Psychology in Maintaining a Growth Mindset and Fostering Continuous Improvement

While data is vital in measuring success, psychology plays an equally important role in sustaining and driving that success over time. The **growth mindset**—a concept popularized by psychologist **Carol Dweck**—is central to fostering continuous improvement within a company.

It is the belief that abilities and intelligence can be developed through dedication and hard work. Leaders who cultivate this mindset in themselves and their teams encourage adaptability, resilience, and a willingness to learn from failure.

A growth mindset can help teams remain focused on **learning and development** rather than being discouraged by setbacks. This psychological approach is especially important during periods of scaling, when businesses are likely to face challenges such as market competition, operational inefficiencies, or customer dissatisfaction. Teams that embrace a growth mindset are more likely to approach problems with curiosity and a desire to solve them, rather than with a fear of failure.

Leadership and Growth Mindset: The role of leadership is crucial in instilling a growth mindset. Leaders can model the behavior they want to see in their teams by embracing challenges, learning from mistakes, and encouraging experimentation. A leader who is transparent about setbacks and demonstrates a willingness to adapt fosters a culture where employees feel safe to innovate and fail forward. This open and supportive environment encourages individuals to take risks and contribute their ideas to drive improvement.

Feedback and Continuous Improvement: For a growth mindset to be effective, feedback is essential. Providing employees with regular, constructive feedback helps them understand where they stand and how they can improve. However, the way feedback is delivered is critical. Leaders should focus on offering feedback that is **actionable**, **specific**, and **growth-oriented**. When feedback is viewed as an opportunity to learn, rather than as a critique, employees are more likely to take initiative and contribute to ongoing improvements.

Creating a culture of **continuous learning** also plays a vital role in fostering a growth mindset. Companies can invest in **training programs**, **mentorship opportunities**, and **professional development** initiatives that provide employees with the skills and knowledge to adapt and grow with the business. This not only benefits the individual employees but also strengthens the company's capacity to innovate and evolve over time.

How to Adjust Business Strategies Based on Evolving Data and Psychological Insights

Business strategies are never static. They must evolve as the business grows, as new challenges arise, and as customer needs change. The ability to adjust strategies based on **data-driven insights** and **psychological factors** is one of the key differentiators between businesses that thrive and those that stagnate.

Data-Driven Adjustments: The continuous monitoring of KPIs and metrics enables businesses to spot emerging trends and identify areas where their strategies may need to be adjusted. For example, if customer churn rates increase or if sales growth begins to plateau, businesses may need to revisit their marketing approach, pricing strategy, or customer engagement efforts. Data analytics allows companies to quickly identify **underperforming areas** and **adjust their approaches** before small issues escalate into bigger problems.

By leveraging **predictive analytics**, businesses can anticipate future trends and make strategic adjustments before they are impacted by those changes. For example, if data indicates that a particular product line is losing popularity, the company can begin shifting its focus toward more profitable or in-demand products.

Psychological Insights for Strategy Adaptation: Psychological factors also play a crucial role in adjusting strategies. For instance, understanding changes in consumer behavior or shifts in cultural attitudes can help businesses refine their messaging and product offerings. When companies scale into new markets, understanding the psychological profiles of their target audience helps tailor their approach to meet local preferences. Businesses can use psychological insights to understand how to communicate the right messages, create customer engagement strategies, and manage the emotions and motivations that drive customer loyalty.

Additionally, when **internal dynamics** evolve, such as team performance or employee morale, businesses must adapt their internal strategies. For example, if employee motivation begins to dip, it might signal the need for a renewed focus on recognition, rewards, or professional development. Understanding the psychological drivers of employee satisfaction and performance can inform decisions on how to best invest in people and maintain a motivated workforce.

As businesses scale, the complexity of the challenges they face increases, and the need to adjust strategies becomes more critical. A flexible, adaptive strategy that incorporates data analytics and psychological insights allows businesses to **navigate uncertainty** and stay ahead of the curve. Companies that embrace continuous improvement through strategic adjustments foster a culture of innovation and resilience, ensuring that they are always moving forward.

Measuring success and fostering continuous improvement are essential to ensuring that a business doesn't just grow but also remains competitive and adaptable over time. The right KPIs and metrics provide the tools necessary to evaluate progress and make informed decisions, while **psychological strategies** like promoting a growth mindset and emphasizing continuous learning ensure that teams remain motivated and resilient through periods of change. By combining data-driven insights with psychological understanding, businesses can effectively adapt their strategies to evolving circumstances, leading to sustained success and growth. Ultimately, those businesses that embrace both measurement and continuous improvement are better positioned to thrive in an increasingly dynamic business landscape.

Chapter 13: Future Trends in Business Growth: The Role of Math and Psychology

As businesses continue to navigate an increasingly complex and competitive landscape, it becomes more critical to understand the shifting dynamics of growth and how emerging technologies, evolving consumer behaviors, and cutting-edge insights will play pivotal roles in shaping business strategies. The future of business growth is not just about maintaining a steady pace of expansion, but about adapting quickly, leveraging new tools, and staying ahead of the curve in a rapidly changing environment. In today's interconnected and fast-paced world, organizations must embrace innovative strategies that combine the analytical power of **mathematical strategies** and the nuanced understanding of **psychological insights**. These combined forces will enable businesses to better understand market trends, predict consumer behavior, and optimize operations with precision and empathy. This chapter delves into the emerging trends in business growth, focusing on the growing significance of **artificial intelligence (AI)**, **big data**, and **behavioral psychology** in shaping future business strategies. Through this exploration, we aim to uncover how businesses can stay ahead of the competition by integrating evolving mathematical and psychological approaches into their operations, ensuring long-term success and sustainability.

Emerging Trends in Business Growth, Technology, and Consumer Behavior

The rapid pace of technological advancements is continuously reshaping the way businesses grow. From automation and artificial intelligence (AI) to shifts in consumer preferences, these innovations present both challenges and opportunities for businesses aiming to maintain sustainable growth. As the world becomes more connected, businesses are experiencing shifts in how products and services are delivered, how customers interact with brands, and even how companies approach scaling.

Technology and Business Growth: Over the next decade, businesses will increasingly rely on **automation** and **AI-powered systems** to streamline operations, enhance customer experiences, and increase overall efficiency. **Machine learning** algorithms will help businesses predict consumer behavior, optimize supply chains, and improve personalized marketing efforts. **AI-powered chatbots** and **virtual assistants** will continue to redefine customer service by offering real-time, personalized responses and resolving issues faster than human representatives could.

Another significant technological trend is the rise of **blockchain technology**, which promises to improve transparency, traceability, and security, especially for industries such as finance, supply chain management, and healthcare. By allowing for secure, decentralized transactions, blockchain could revolutionize business models and the way companies scale, particularly in terms of **international transactions** and **cross-border trade**.

Consumer Behavior Shifts: As technology continues to evolve, so too does consumer behavior. The digital transformation has drastically altered how customers interact with businesses. **Personalization** has become key; consumers expect tailored experiences and product recommendations based on their preferences, purchase history, and browsing patterns. The increasing use of **mobile apps** and **social media** platforms for discovering and purchasing products means that businesses must adopt **omnichannel strategies** to reach customers across multiple touchpoints.

Additionally, consumers are becoming more **conscious of social and environmental factors**, and companies are expected to align their growth strategies with **sustainability** and **corporate social responsibility (CSR)** efforts. Businesses that prioritize ethical practices, environmental sustainability, and community engagement are likely to attract a more loyal customer base, particularly among younger generations who are more socially aware.

The future of business growth is also significantly influenced by **consumer empowerment**. With the rise of **online reviews**, **social media influencers**, and **crowdsourced platforms**, customers now have more influence over business decisions and brand perceptions than ever before. Businesses will need to respond to customer feedback quickly and authentically, adapting to their needs to foster loyalty and maintain positive brand reputations.

The Increasing Role of Artificial Intelligence, Big Data, and Behavioral Insights in Shaping Business Strategies

In the future, the combination of **artificial intelligence**, **big data**, and **behavioral psychology** will be the driving force behind business strategies. The ability to analyze vast amounts of data and gain actionable insights will enable businesses to predict market trends, understand customer preferences, and make more informed decisions. These technologies will help businesses create highly targeted strategies that are personalized, efficient, and responsive to ever-changing consumer demands.

Artificial Intelligence (AI) and Business Strategy: AI is already transforming business operations, but its potential to reshape industries in the future is vast. With AI, companies can analyze massive datasets in real time, allowing them to identify trends, forecast demand, and automate decision-making processes. For example, AI-driven recommendation engines in e-commerce platforms use data on customers' previous purchases and browsing habits to suggest products, thus enhancing the customer experience and increasing sales.

AI is also being used to optimize supply chains, ensuring that businesses maintain optimal stock levels and reduce waste. **Predictive analytics** powered by AI can help companies anticipate demand fluctuations, adjust inventory accordingly, and reduce operational costs. Moreover, AI's ability to personalize marketing campaigns based on individual behaviors means that businesses can create highly effective targeted marketing strategies that resonate with consumers on a deeper level.

Big Data and Data-Driven Decision Making: Big data has revolutionized how businesses make decisions. The ability to collect and analyze data from various sources—social media, customer feedback, sales reports, and more—provides companies with a comprehensive view of their operations and market conditions. By analyzing this data, businesses can gain insights into everything from **customer sentiment** to **market trends**, helping them adjust strategies accordingly.

Data analytics tools and **predictive modeling** enable businesses to anticipate shifts in consumer behavior and demand. For instance, companies can use big data to analyze purchasing patterns and tailor their inventory, marketing, and product development strategies based on consumer behavior. Additionally, businesses can use big data to enhance customer experience through **dynamic pricing models** and **real-time personalization**.

Behavioral Insights in Shaping Strategy: Understanding the psychology of consumers is essential in shaping effective business strategies. **Behavioral economics**—the study of how psychological, social, and emotional factors influence consumer decisions—is increasingly becoming a vital aspect of business planning. Businesses are using behavioral insights to optimize their product offerings, pricing strategies, and customer experiences by considering the cognitive biases that influence purchasing decisions.

For example, businesses use **anchoring** (the tendency for customers to rely heavily on the first piece of information they encounter) to price products more effectively. **Loss aversion** (the concept that people fear losing more than they desire to gain) can be used in sales promotions and offers. Understanding these psychological biases allows businesses to optimize their marketing messages, pricing models, and customer interactions to increase sales and customer loyalty.

Moreover, businesses are increasingly focusing on **emotional engagement** with customers, using psychological principles to build stronger emotional connections with their audience. For instance, using **storytelling** or **visual branding** can evoke emotions that resonate with consumers and foster long-term loyalty. Integrating these behavioral insights into the overall strategy helps businesses appeal to customers on a deeper, more personal level, encouraging repeat purchases and advocacy.

How Businesses Can Stay Ahead by Integrating Evolving Mathematical and Psychological Approaches

To stay ahead in a rapidly changing business environment, companies must adopt an integrated approach that combines **evolving mathematical models** with **psychological insights**. The fusion of these two domains can help businesses not only measure success and drive growth but also adapt quickly to changing conditions and consumer demands.

Mathematical Models and Predictive Analytics: Mathematical approaches, particularly predictive modeling and **machine learning algorithms**, will continue to evolve and play a central role in business strategy. By analyzing vast amounts of data and applying advanced statistical techniques, businesses can gain more accurate forecasts of market trends and consumer behavior. As AI and data analytics technologies become more sophisticated, businesses will be able to make data-driven decisions with greater precision and efficiency.

For example, businesses will increasingly rely on predictive algorithms to optimize their product development cycles, customer service, and inventory management. Additionally, **sentiment analysis** powered by AI will allow businesses to track consumer feelings toward their products or brand, helping them adjust their marketing efforts and improve customer experiences in real time.

Psychological Insights and Emotional Intelligence: While data and analytics are critical for making informed decisions, the human element remains central to building strong customer relationships. Businesses that integrate psychological insights into their strategies will have a competitive advantage in fostering emotional connections with their customers. Developing **empathy** and **emotional intelligence** within business teams will be key in adapting to evolving consumer needs and creating customer-centric strategies.

Psychological approaches can also inform how businesses build trust, engage with consumers on social media, and address customer complaints. Companies that recognize the emotional drivers of consumer behavior can create **experiences** that go beyond functional product offerings and foster lasting customer loyalty.

Integrating Both Approaches: The future of business growth lies in the ability to integrate data-driven, mathematical approaches with psychological understanding. Businesses that harness the power of **AI**, **big data**, and **behavioral psychology** will be able to optimize customer acquisition, retention, and satisfaction. Moreover, adopting a **holistic strategy** that takes both data and psychology into account allows businesses to craft personalized, engaging experiences that resonate with consumers on multiple levels.

As companies continue to expand into new markets, adopt emerging technologies, and deal with evolving consumer demands, staying ahead requires more than just keeping up with trends. It requires a dynamic approach that blends the precision of data analytics with the nuance of human behavior.

The future of business growth will be shaped by the intersection of emerging technologies, evolving consumer behavior, and the growing understanding of psychological principles. The combination of **artificial intelligence**, **big data**, and **behavioral insights** will empower businesses to not only predict trends but also adapt to changes in the market with precision and empathy. To stay ahead in this dynamic environment, businesses must continually integrate both **mathematical models** and **psychological strategies** into their overall growth framework. By embracing these evolving approaches, companies will be better equipped to navigate the complexities of the future and drive long-term, sustainable growth.

Chapter 14: Conclusion: Building a Sustainable Growth Strategy for the Long Term

As we reach the conclusion of this exploration into the intricacies of business growth, it is important to reflect on the core principles that have been presented throughout the book. This chapter serves as a recap of the key takeaways, emphasizing how **math**, **data**, and **psychology** work together synergistically to drive long-term business success. We will also discuss the final thoughts on constructing a growth-oriented business model that not only thrives in today's fast-paced environment but is adaptable to the future. Finally, this chapter encourages readers to apply the principles outlined in the previous chapters, providing the tools and frameworks necessary to build a sustainable and growth-oriented strategy for the future.

Recap of Key Takeaways: How Math, Data, and Psychology Work Together to Fuel Business Expansion

The journey through this book has emphasized that growth is not merely a product of chance, but the outcome of careful planning, strategic thinking, and the application of **mathematical rigor**, **data-driven insights**, and **psychological understanding**. These three elements—**math**, **data**, and **psychology**—work together in ways that enable businesses to optimize their operations, predict trends, and understand consumer behavior at a deep level. By integrating these disciplines, businesses are empowered to create innovative solutions, respond quickly to market changes, and foster stronger relationships with customers.

Mathematics, specifically in the form of **predictive analytics** and **modeling**, helps businesses forecast trends, optimize processes, and make data-backed decisions. Whether it's using algorithms to predict consumer purchasing behavior or applying statistical models to analyze market performance, math provides the foundation for informed, evidence-based decision-making.

Data plays a crucial role in the modern business world, offering insights into everything from customer preferences to operational inefficiencies. The continuous collection and analysis of big data allow businesses to make **real-time adjustments**, enhance customer engagement, and streamline their processes. Businesses that harness the power of data are able to identify emerging opportunities and mitigate risks in ways that were once impossible.

Finally, **psychology** adds the human element to business strategies. By understanding the **emotional drivers** behind consumer decisions and leveraging principles of **behavioral economics**, businesses can create personalized experiences that resonate deeply with their customers. Whether through **personalized marketing** or **strategic product development**, psychological insights enable companies to connect with customers on a more personal and emotional level, creating long-lasting loyalty and engagement.

The harmonious combination of these three pillars—mathematics, data, and psychology—forms a powerful toolkit for businesses aiming to grow and adapt in an ever-changing market.

Final Thoughts on Building a Growth-Oriented Business Model That Adapts to the Future

As we look to the future, the most successful businesses will be those that are able to build flexible, **growth-oriented business models** capable of adapting to change. The key to sustainable growth lies in a business's ability to continuously innovate, respond to shifts in consumer preferences, and navigate disruptions in the market. No business model is immune to change, but businesses that embrace adaptability will be best positioned for long-term success.

A growth-oriented business model is one that **values innovation**, **prioritizes customer experience**, and is built on a foundation of **data-driven decision-making**. In the future, businesses that succeed will not only rely on **math** and **data** to analyze trends and optimize performance but will also deeply understand the **psychology** of their customers. Understanding customer needs, motivations, and emotional triggers will allow businesses to develop products, services, and experiences that meet both **practical needs** and **emotional desires**, which in turn will foster stronger, more loyal customer relationships.

Businesses must also embrace a mindset of **continuous improvement**. The process of growth doesn't end with the initial success—sustainable growth comes from the willingness to **iterate**, **experiment**, and **adapt**. Leveraging insights from **data analytics** and **psychological understanding**, businesses should consistently assess their performance, identify areas of opportunity, and make adjustments as necessary.

Furthermore, **long-term sustainability** requires that businesses develop **resilient systems** that can weather the inevitable challenges that come with growth, such as market fluctuations, technological disruptions, and evolving consumer behaviors. The most successful businesses will be those that remain committed to **scalability**, ensuring that their models can expand without sacrificing quality or customer satisfaction.

Encouraging Readers to Apply the Principles Outlined in the Book for Long-Term Success

As we conclude this journey through the principles of growth and expansion, I encourage readers to take the insights shared in this book and put them into action. The concepts outlined—whether it's applying **predictive models** to anticipate market trends, using **data-driven insights** to improve customer engagement, or leveraging **psychological principles** to enhance marketing efforts—are tools that can significantly improve business performance when put into practice.

Start by assessing your current business model and identify areas where you can integrate the **three pillars** discussed: **math**, **data**, and **psychology**. Incorporate mathematical models to make smarter, more informed decisions. Invest in systems that allow you to collect and analyze **data** in real-time to understand your customers and optimize your operations. Explore **psychological insights** to enhance your customer interactions, create more personalized experiences, and build lasting relationships.

Keep in mind that building a sustainable growth strategy requires patience and persistence. It's not about expecting overnight success, but rather creating a plan that positions your business for **long-term expansion**. Adopt a growth mindset—one that values **experimentation**, **learning from failures**, and **adapting** to new information as it becomes available.

The world of business is evolving at an unprecedented pace, and those who can harness the power of **data**, **technology**, and **psychological insights** will be the ones who lead the charge. The principles outlined in this book are not just theoretical; they are practical tools that, when applied effectively, can help your business grow, evolve, and thrive for years to come.

Building a **sustainable growth strategy** for the long term requires a deep understanding of the forces that drive business success—**mathematics**, **data**, and **psychology**. These elements, when used in tandem, allow businesses to navigate the complexities of the modern market, anticipate change, and craft strategies that foster long-term expansion. By staying agile, leveraging insights, and adapting to future trends, businesses can create models that are both resilient and scalable.

As a final thought, remember that growth is not a one-time event, but a continuous journey. It requires ongoing effort, strategic adaptation, and a willingness to evolve. With the right tools and insights, your business can not only keep pace with change but actively shape the future of growth. So, as you move forward, apply the lessons from this book and position yourself for sustained success.

References:

1. **Fader, P. S., & Hardie, B. G. (2013).** *Marketing Metrics: The Definitive Guide to Measuring Marketing Performance.* Pearson Education.
   - This book provides insights into how data and mathematical models are essential for understanding and improving business performance through marketing metrics.
2. **Kahneman, D. (2011).** *Thinking, Fast and Slow.* Farrar, Straus and Giroux.
   - Nobel laureate Daniel Kahneman's work on behavioral economics and cognitive biases provides the foundation for understanding consumer behavior and decision-making in business.
3. **Brynjolfsson, E., & McAfee, A. (2014).** *The Second Machine Age: Work, Progress, and Prosperity in a Time of Brilliant Technologies.* W.W. Norton & Company.
   - This book examines the role of machine learning and AI in business transformation, particularly for scaling operations and driving growth.
4. **Tversky, A., & Kahneman, D. (1974).** *Judgment under Uncertainty: Heuristics and Biases.* Science, 185(4157), 1124-1131.
   - A classic paper on the psychological biases that influence decision-making, offering critical insights for applying behavioral economics in business growth strategies.
5. **Porter, M. E. (1990).** *The Competitive Advantage of Nations.* Free Press.
   - Porter's work on competitive strategy and growth highlights the importance of strategic decision-making models and how they can be applied to market expansion.
6. **Davenport, T. H., & Harris, J. G. (2007).** *Competing on Analytics: The New Science of Winning.* Harvard Business Review Press.
   - This book explores the growing importance of data-driven decision-making and how businesses can leverage analytics to achieve competitive advantages and sustainable growth.
7. **Chadeau-Hyam, M., et al. (2019).** *The Role of Mathematical Modeling in Business and Economics.* Mathematical Models in Economics and Business, 2(1), 13-29.

- This academic article delves into the integration of mathematical models in business practices, particularly their application in forecasting and scaling.

8. **Cialdini, R. B. (2009).** *Influence: The Psychology of Persuasion.* Harper Business.
   - Robert Cialdini's book on the psychology of influence and persuasion is a key resource for understanding consumer behavior and how to craft marketing strategies that drive business growth.

FSC
www.fsc.org
MIX
Papier | Fördert
gute Waldnutzung
FSC® C083411

Kolibri 360 GmbH
Ferdinand-Jühlke-Straße 7
99095 Erfurt, Deutschland
produktsicherheit@kolibri360.de

Druck:
CPI Druckdienstleistungen GmbH
im Auftrag der
Kolibri 360 GmbH
Ferdinand-Jühlke-Str. 7
99095 Erfurt